The ARRL
Ham Radio License Manual

All you need to become an
Amateur Radio Operator

1st Edition

By Ward Silver, NØAX

Production Staff

David Pingree, N1NAS, Senior Technical Illustrator:

Jodi Morin, KA1JPA, Assistant Production Supervisor: Layout

Kathy Ford, Proofreader

Sue Fagan, Graphic Design Supervisor: Cover Design

Michelle Bloom, WB1ENT, Production Supervisor: Layout

ARRL *The national association for* **AMATEUR RADIO**

225 Main Street, Newington, CT 06111-1494

This book may be used for Technician license exams given beginning July 1, 2006.

We strive to produce books without errors. Sometimes mistakes do occur, however. When we become aware of problems in our books (other than obvious typographical errors that should not cause our readers any problems) we post an Adobe Portable Document Format (PDF) file on **ARRLWeb**. If you think you have found an error, please check **www.arrl.org/notes** and **www.arrl.org/hrlm** for corrections. If you don't find a correction there, please let us know, either using the Feedback Form in the back of this book or by sending an e-mail to **pubsfdbk@arrl.org**.

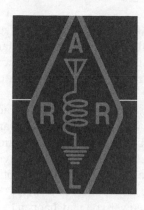

Contents

Preface

Welcome to the exciting world of Amateur Radio! You are about to join nearly 700,000 licensed Amateur Radio operators in the United States and nearly three million people around the world who call themselves "hams." Hams are found in virtually every country in the world. They have earned the special privilege of being able to communicate directly with one another, by radio, without regard to the geographic and political barriers that so often limit our understanding of the world.

There are many important reasons why governments allow Amateur Radio operators to use valuable radio frequencies for personal communications. As a licensed Amateur Radio operator, you will become part of a large group of trained communicators and electronics technicians. You will be an important emergency communications resource for your neighbors and fellow citizens. Who knows when you may find yourself in the situation of having the only communications link outside your neighborhood? Whether you are caught in a flood, earthquake or other natural disaster or answering a call from someone else, you can provide the knowledge and resources to help.

Whether across town or across the sea, hams are always looking for new friends. So wherever you may happen to be, you are probably near someone — perhaps a whole club — who would be glad to help you get started. If you need help contacting hams, instructors, Volunteer Examiners or clubs in your area, contact us here at ARRL Headquarters. We'll help you get in touch with someone near you. (See the contact information at the bottom of this page.)

When you pass that exam and enter the exciting world of Amateur Radio, you'll find plenty of activity to keep you busy. You'll also find plenty of friendly folks who are anxious to help you get started. Amateur Radio has many interesting areas to explore. You may be interested in one particular aspect of the hobby now, but be willing to try something new occasionally. You'll discover a world of unlimited potential!

Most of the active radio amateurs in the United States are members of ARRL. The hams' own organization since 1914, ARRL is truly the national association for Amateur Radio. We provide training materials and other services, and represent our members nationally and internationally. *The ARRL Ham Radio License Manual* is just one of the many ARRL publications for all levels and interests in Amateur Radio. You don't need a ham license to join. If you're interested in ham radio, we're interested in you. It's as simple as that!

David Sumner, K1ZZ
Executive Vice President
Newington, Connecticut
March 2006

The ARRL Ham Radio License Manual   ON THE WEB

WWW.ARRL.ORG/HRLM

Visit *The ARRL Ham Radio License Manual* home on the Web for additional resources as you prepare for your first Amateur Radio license.

New Ham Desk
ARRL Headquarters
225 Main Street
Newington, CT 06111-1494
(860) 594-0200

Prospective new amateurs call:
800-32-NEW-HAM (800-326-3942)
You can also contact us via e-mail: **newham@arrl.org**
or check out **ARRLWeb: www.arrl.org/**

The Adventure Begins!

Congratulations! You've taken your first step into a hobby—and a service—that knows no limits. Amateur Radio is a worldwide network of people from various cultures, united by a common love of wireless communication. Amateur Radio is as old as radio itself, and its future is no less fantastic than its past.

For most people, Amateur Radio is a lifelong pursuit. We want to make sure you get a good start, which is why we've published this book. But first, who are "we"?

ARRL The National association for Amateur Radio: What's in it for You?

♦ **Help for New Hams**: Are you a beginning ham looking for help in getting started in your new hobby? The hams at ARRL HQ in Newington, Connecticut, will be glad to assist you. Call 800-32-NEWHAM. ARRL maintains a computer data base of ham clubs and ham radio "helpers" from across the country who've told us they're interested in helping beginning hams. There are probably several clubs in your area! Contact us for more information.

♦ **Licensing Classes**: If you're going to become a ham, you'll need to find a local license exam opportunity sooner or later. ARRL Registered Instructors teach licensing classes all around the country, and ARRL-sponsored Volunteer Examiners are right there to administer your exams. To find the locations and dates of Amateur Radio Licensing classes and test sessions in your area, call the New Ham Desk at 800-32-NEWHAM.

♦ **Clubs**: As a beginning ham, one of the best moves you can make is to join a local ham club. Whether you join an all-around group or a special-interest club (repeaters, DXing, and so on), you'll make new friends, have a lot of fun, and you can tap into a ready reserve of ham radio knowledge and experience. To find the ham clubs in your area, call HQ's New Ham Desk at 800-32-NEWHAM.

♦ **Technical Information Service**: Do you have a question of a technical nature? (What new ham doesn't?) Contact the Technical Information Service (TIS) at HQ. Our resident technical experts will help you over the phone, send you specific information on your question (antennas, interference and so on) or refer you to your local ARRL Technical Coordinator or Technical Specialist. It's expert information—and it doesn't cost Members an extra cent!

♦ **Regulatory Information**: Need help with a thorny antenna zoning problem? Having trouble understanding an FCC regulation? Vacationing in a faraway place and want to know how to get permission to operate your ham radio there? HQ's Regulatory Information Branch has the answers you need!

♦ **Operating Awards**: Like to collect "wallpaper"? The ARRL sponsors a wide variety of certificates and Amateur Radio achievement awards. For information on awards you can qualify for, contact the Membership Services Department at HQ.

♦ **Equipment Insurance**: When it comes to protecting their Amateur Radio equipment investments, ARRL Members travel First Class. A. H. Wohlers Company provides ARRL's "all-risk" equipment insurance plan. (It can protect your ham radio computer, too.) It's comprehensive and cost effective, and it's available only to ARRL Members. Why worry about losing your valuable radio equipment when you can protect it for only a few dollars a year?

♦ **Amateur Radio Emergency Service**: If you're interested in providing public service and emergency communications for your community, you can join more than 25,000 other hams who have registered their communications capabilities with local Emergency Coordinators. Your EC will call on you and other ARES members for vital assistance if disaster should strike your community. Contact the Field Services Department at HQ for information.

♦ **Audio-Visual Programs**: Need a program for your next ham club meeting, informal get-together or public display? ARRL Affiliated clubs can buy tapes from ARRL's video programs. There are many to choose from, including popular titles on Amateur Radio's role in Operation Desert Storm and space shuttle activities. Contact Field & Educational Services at HQ for a complete list or to order a tape.

♦ **Blind, Disabled Ham Help**: For a list of available resources and information on the Courage HANDI-HAM System, contact the ARRL Program for the Disabled at HQ.

With your membership you also receive the monthly journal *QST*. Each 200-page issue is packed with valuable information you can use—including a special *QST Workbench* section with information of special interest to new hams! And *QST* Product Reviews are the most respected source of information to help you get the most for your Amateur Radio equipment dollar. (For many hams, *QST* alone is worth far more than the cost of ARRL membership.)

The ARRL also publishes newsletters and dozens of books covering all aspects of Amateur Radio. Our Headquarters station, W1AW, transmits bulletins of interest to radio amateurs and Morse Code practice sessions.

When it comes to representing Amateur Radio's best interests in our nation's capital, ARRL's team in Washington, DC, is constantly working with the FCC, Congress and industry to protect and foster your privileges as a ham operator.

Regardless of your Amateur Radio interests, ARRL Membership is relevant and important. We will be happy to welcome you as a Member. Use the Invitation to Membership on the next page to **join today**. And don't hesitate to contact us if you have any questions!

How to Use this Book

We designed *The ARRL Ham Radio License Manual* both for self study and for classroom use. An interested student will find this book complete, readable and easy to understand. Read carefully, and test yourself often as you study. Before you know it, you'll be ready to pass that exam!

Why deprive yourself of the company of fellow beginners and the expertise of those "old-timers" in your hometown, though? *The ARRL Ham Radio License Manual* goes hand-in-hand with a very effective ARRL-sponsored training program run by over 2000 volunteer instructors throughout the United States. If you would like to find out about a local class, just contact the New Ham Desk at ARRL Headquarters—we'll be happy to assist. (Our phone number and e-mail address were listed earlier.)

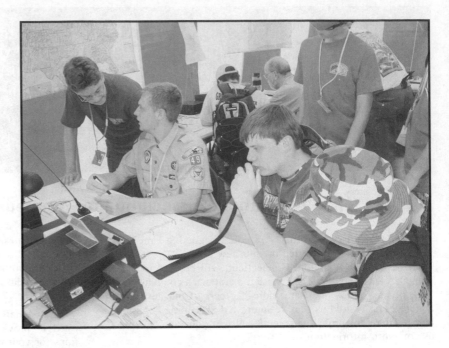

Hams are very social animals who derive a great deal of pleasure from helping a newcomer along the way. The most effective learning situation is often the one you share with others. There are knowledgeable people to turn to when you have a question or problem. Fellow students can quiz one another on basic electronics concepts and operating practices. It doesn't matter if you're studying on your own or joining a class, though. Use *The ARRL Ham Radio License Manual* to study for your Technician exam and you'll be on the air in no time at all.

Bite-Sized Information

The ARRL Ham Radio License Manual presents the material in easily digested and well-defined "bite-sized" sections. You will be directed to turn to the

question pool at the back of the book as you complete a section of the material.

This review will help you determine if you're ready to move on. It will also highlight those areas where you need a little more study. In addition, this approach takes you through the entire question pool. By the time you complete the book, you will be familiar with all the questions used to make up your test. Please take the time to follow these instructions. Believe us, it's better to learn the material correctly the first time than to rush ahead, ignoring weak areas and unresolved questions.

Every page presents information you'll need to pass the exam and become an effective operator. Pay attention to diagrams, photographs, sketches and captions; they contain a wealth of information you should know. You'll also find a few anecdotes and "mini-articles" (called Sidebars) that will help put the tradition of Amateur Radio in perspective. Our roots go back to the beginning of the 20th Century, and our community service continues even as you read this.

Start at the beginning. Study the material presented in this book and follow the instructions to review the exam questions. You'll cover small sections of the text and a few questions at a time. This is the best way to determine how well you understand the most important points. The text explains the theory in straightforward terms, so you shouldn't have any problems. Review the related sections if you have any difficulty, and then go over the questions again. In this way, you will soon be ready for your exam. Before you know it, you'll be on the air!

Don't be afraid to ask for help if you don't

understand something, though. If you're participating in a class taught by an ARRL-registered instructor, you'll have the chance to ask the experts for help. Ask your instructor about anything you are having difficulty with. You may also find it helpful to discuss the material with your fellow students. If you're not in a class and run into snags, don't despair! The folks at the New Ham Desk at ARRL Headquarters will be happy to put you in touch with an Amateur Radio operator in your area who can help answer your questions.

Most people learn more when they are actively involved in the learning process. Turning to the questions and answers when directed in the text helps you be actively involved. Check your answers after you study each group of questions, and review the appropriate text for any questions you get wrong. Paper clips make excellent place markers to help you keep your spot in the text and the question pool.

Before You Take Your Test

If you need help in locating someone to administer a Technician test, or for a schedule of Volunteer Examiner test sessions in your area, write to the ARRL/VEC Office, ARRL Headquarters, 225 Main St, Newington, CT 06111-1494. If you need help locating a ham, instructor or club, contact the New Ham Desk at ARRL Headquarters as listed earlier. We can put you in touch with examiners, instructors and clubs in your area! *ARRLWeb* also includes exam and club search sections to help you find local hams on line. Go to **www.arrl.org** and follow the site index to "clubs" or "exams" for the information you need.

Give *The ARRL Ham Radio License Manual* a chance to guide you the way it was intended — by following these instructions. You'll soon be joining us on the air. Each of us at the ARRL Headquarters and the entire ARRL membership wishes you the very best of success. We are all looking forward to that day in the not-too-distant future when we hear your signal on the ham bands. 73 (best regards) and good luck!

Chapter 1

Welcome to Amateur Radio

This first section of the book will explain:

- **What the Technician license is**
- **How to study with this book**
- **Where you can find other hams**
- **What makes Amateur Radio unique**
- **Why the FCC makes the rules**
- **What activities you'll find in Amateur Radio**

Ready? Set? Go!

Welcome to *The ARRL Ham Radio License Manual*, the most popular introduction to Amateur Radio of all! You're in good company—there are thousands of other folks getting ready to join the ranks of "ham" radio operators. In this study guide, not only will you learn enough to pass your Technician license exam, you'll also learn what ham radio is all about and how to jump right in once you're ready to go on the air.

1.1 Getting your HAM RADIO License

THE TECHNICIAN LICENSE

The Technician license is the first license for most newcomers to ham radio. There are more Technician licensees than of any other class, nearly 50% of all hams. No Morse code exam is required to earn a Technician-class license but you can learn and use "the code" as much or as little as you want. In the meantime, you'll be able to communicate with thousands of other hams in any of the many ways amateurs use the airwaves.

Once you gain some experience, you'll be ready to upgrade your license to General class and beyond to the top of the line Amateur Extra class. These licensees gain more privileges on the traditional HF or "short wave" bands of Amateur Radio. They all started just like you, taking the basic exams and getting on the air.

OBTAINING A LICENSE

The first step is in your hands right now! To get your license, you'll need to pass a 35-question, multiple-choice written exam on the rules of ham radio, simple operating procedures, and basic electronics. Carefully study *The ARRL Ham Radio License Manual* until you can answer the exam questions shown in the text.

You can study on your own or you can enroll in a licensing

Dennis Motschenbacher, K7BV, enjoys a ham contest while visiting the Dominican Republic

class. Log on to the ARRL Web site at **www.arrl.org** and click "Classes." You'll be redirected to a Web page where you can search for classes being held near you. If you prefer, contact the ARRL's New Ham Desk directly at the addresses in this book. There are more than 2000 instructors throughout the United States that are part of the ARRL-sponsored training program. By joining a class, you can take advantage of the experience of these ham radio experts and learn in the company of other students.

When you are ready to take your exam, it will be time to locate an exam session. If you're part of a study class, the instructor will make the necessary arrangements. For solo students, you can find an exam session by clicking "Exams" on the ARRL home page mentioned above. You'll be directed to the Licensing Classes search page to find exam sessions near you, including complete contact information. All Amateur Radio exams are given by ham radio operators acting as volunteer examiners.

After you pass your exam, the examiners will give you a *Certificate of Successful Completion of Examination* (CSCE) that documents your achievement. They will also file all of the necessary paperwork so that your license will be granted by the Federal Communications Commisssion (FCC). In a few days, you will be able see your new call sign in the FCC's database via the ARRL's Web site! Congratulations—you're on the air! Later, you'll receive a paper license by mail.

HOW TO USE THIS BOOK

The ARRL Ham Radio License Manual has seven sections, starting with the basics of ham radio equipment and the fundamentals of how radio works. You'll then learn how to put together a simple station. The next sections show you how to follow the rules and regulations of ham radio. The last section is all about ham radio safety. Each section has practical examples and information you can use for reference later. As you learn about each topic, questions from the license exam are provided so that you can be sure you understand the material.

If you are taking a licensing class, help your instructors by letting them know about areas in which you need help. They want you to learn as thoroughly and quickly as possible, so don't hold back with your questions. Similarly, if you find the material particularly clear or helpful, tell them that, too, so it can be used in the next class!

In the back of the book you'll find an Appendix including the complete set of exam questions and answers. There is also a large glossary of ham radio words, useful tables and lists, a detailed index, and a section of advertisements from some of Amateur Radio's best-known vendors of equipment and supplies.

An On-Line Mentor

The ARRL's "Welcome to Ham Radio" Web page at **www.arrl.org/hamradio.html** has several links right at the top of the page for licensing classes, study materials, background information, and more. You'll also want to bookmark the The ARRL Ham Radio License Manual Web page, **www.arrl.org/hrlm** because of the many references to it in this book. More detailed technical information about many of the topics here can be found at the "First Steps in Radio" page, **www.arrl.org/tis/info/frsteps.html**.

WHAT WE ASSUME ABOUT YOU

The only thing you'll really need to succeed is a strong interest in Amateur Radio and a willingness to learn. You don't have to be a technical guru or an expert operator to get your license! As you progress through the material, you'll encounter some basic science about radio and electricity. There will be a simple bit of math here and there. When we get to the rules and regulations you'll have to learn some new words and maybe memorize a few numbers. That's it! It will help if you have regular access to the Internet. (It's not necessary to have a high-speed connection, dial-up will do just fine.) You should have a simple calculator, which you'll also be allowed to use during the license exam.

ADVANCED STUDENTS

Perhaps you've used other types of radios, such as Citizen's Band or a business band radio at work. You might have received some technical or operator training. If so, we suggest that you jump to the Question Pool in the Appendix and review the questions and answers. If you find it easy to answer some of the questions correctly, you can skim or skip the corresponding sections of the book. Each question is cross-referenced to the section in *The ARRL Ham Radio License Manual* that provides background on that topic. Regardless of your background, be sure to review the section on Operating Legally since ham radio procedures and conventions may be different than what you're used to.

SELF-STUDY OR CLASSROOM STUDENTS

The ARRL Ham Radio License Manual can be used either by an individual student, studying on his or her own, or as part of a licensing class taught by an instructor. If you're part of a class, the instructor will guide you through the book, section by section. The solo student can move at any pace and in any convenient order. You'll find that having a buddy to study with makes learning the material more fun as you help each other over the rough spots.

Don't hesistate to ask for help! Your instructor can provide information on anything you find difficult. Classroom students may find asking their fellow students to be helpful. If you're studying on your own, there are resources for you, too! If you can't find the answer via the *The ARRL Ham Radio License Manual* Web page listed below, email your question to the ARRL's New Ham Desk, **newham@arrl.org**. The ARRL's experts will answer directly or connect you with another ham operator that can answer your questions.

Whether you take a class or decide on self-study, we encourage each student to take advantage of the many on-line resources noted in the text. *The ARRL Ham Radio License Manual* Web page, **www.arrl.org/hrlm**, contains many useful links, organized in sections just like this book.

USING THE QUESTION POOL

Resist the temptation to just memorize the answers. Doing so leaves you without the real understanding that will make ham radio enjoyable and useful. *The ARRL Ham Radio License Manual* covers every one of the exam questions, so you can be sure you're ready at exam time.

When using the Question Pool, cover or fold over the answers at the edge of the page to be sure you really do understand the question. Each question also includes a cross-reference back to the section of the book that covers that topic. If you don't completely understand the question or answer, please go back and review that section. The ARRL's condensed guide *Tech Q&A* also provides explanations for each one of the exam questions.

ON-LINE EXAMS

When you feel like you're nearly ready for the actual exam you can get some good practice by taking one of the on-line Amateur Radio exams. These Web sites use the same questions and answers to construct an exam with the same number and variety of questions that you'll encounter on exam day. The exams are free and you can take them over and over again in complete privacy. Links to on-line exams can be found on the ARRL's *The ARRL Ham Radio License Manual* Web page.

These exams are quite realistic and you get quick feedback about the questions you missed. When you find yourself passing the on-line exams by a comfortable margin, you'll be ready for the real thing! A note of caution, be sure that the questions used are current—the Technician question pool is completely rewritten every three years. The set of questions put in place in July of 2006 will be replaced in 2009. (Other license class question pools will expire in 2007 and 2008.)

CONVENTIONS & RESOURCES

Throughout your studies keep a sharp eye out for words in *italics*. These words are important so be sure you understand them. Be sure to make use of the extensive Glossary in the back of the book. Another thing to look for are the addresses or URLs for Web resources, such as **www.arrl.org/hrlm**. By browsing these Web pages while you're studying *The ARRL Ham Radio License Manual*, you will accelerate and broaden your understanding.

Throughout the book, there are also many short sidebars that present topics related to the subject you're studying. These sidebars may just tell an interesting story or they might tackle a subject that needs its own space in the book. The information in sidebars helps you understand how the information you're studying relates to ham radio in general. You'll enjoy reading about the history and traditions of ham radio, too.

Use the *Ham Radio License Manual* Web page as an on-line reference while you study. Many of the Web resources we provide in this book link through the *Ham Radio License Manual* Web page. The Web page lists other resources organized by section and chapter to follow the book. Browse these links for extra information about the topics in this book. For a focused discussion on each exam question, pick up a copy of the *ARRL's Tech Q&A*. Every question is there, with the correct answer and a short paragraph or two on the topic.

Books to Help You Learn

As you study the material on the licensing exam, you will have lots of other questions about the hows and whys of Amateur Radio. The following references, available from your local bookstore or the ARRL (**www.arrl.org/catalog**) will help "fill in the blanks" and give you a broader picture of the hobby:

• *Ham Radio for Dummies* by Ward Silver—written for the prospective and new ham that will be wondering "What do I do now?." The book supplements the information in study guides with an informal, friendly approach to the hobby.

• *ARRL Operating Manual*—in-depth chapters on the most popular ham radio activities. Learn about nets, award programs, DXing and more, with a healthy set of reference tables and maps.

• *Understanding Basic Electronics*—for students that want more technical background about radio and electronics. The book covers the fundamentals of electricity and electronics that are the foundation of all radio.

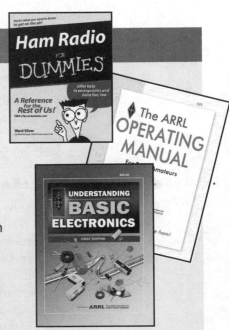

TESTING PROCESS

The final step is to find a *test session*. If you're in a licensing class, the instructor will help you find and register for a session. Otherwise, you can find a test session by using the ARRL's Web page for finding exams, **www.arrl.org/examsearch**. If you can register for the test session in advance, do so. Other sessions, such as those at hamfests or conventions, are available to anyone that shows up or *walk-ins*. You may have to wait for an available space though, so go early!

Bring two forms of identification including at least one photo ID, such as a driver's license, passport, or employer's identity card. Also know your Social Security Number (SSN). You can bring pencils or pens and a calculator, but any kind of computer or on-line devices are prohibited. (If you have a disability and need these devices to take the exam, contact the session sponsor ahead of time.) Once you're signed in, you pay the test fee (currently $14) and get ready.

Amateur Radio licensing test sessions are administered by volunteer examiners—hams just like you will be. They grade the exams, help you fill out the necessary forms, and take care of all the paperwork for your ham radio license.

The Technician test takes from 15 to 45 minutes. You will be given a question sheet and an answer sheet. As you answer each question, you'll mark a box on the answer sheet. Once you've answered all 35 questions, the volunteer examiners (VE) will grade and verify your test results. Assuming you've passed (congratulations!) you'll fill out a Certificate of Successful Completion of Examination (CSCE) and a NCVEC Form 605. The exam organizers will submit your results to the FCC while you keep the CSCE as evidence that you've passed your Technician test. As soon as your name and call sign appear in the FCC's database of licensees, typically a week to 10 days later, you can start transmitting!

If you don't pass, don't be discouraged! You might be able to take another version of the test right then and there if the session organizers can accommodate you. Even if you decide to try again later, you now know just how the test session feels—you'll be more relaxed and ready next time. The ham bands are full of hams that took their tests more than once before passing. You'll be in good company!

TIME TO GET STARTED!

By following these instructions and carefully studying the material in this book, soon you'll be joining the rest of us on the air! Each of us at the ARRL Headquarters and every ARRL member looks forward to the day when your signals join ours on the ham bands. '73' (best regards) and good luck!

1.2 Amateur Radio Clubs and Organizations

Your ham radio support group comes in many forms—a fellow student or classmate, a nearby ham, a club, or even a nationwide organization. All of them are resources for you, not only during your studies for the licensing exam, but also after you have your call sign and are learning how to be a ham. Helping newcomers is one of ham radio's oldest traditions. After all, we are "amateurs" together. Nearly every ham has mentored or *Elmered* another ham at one time or another. You'll be amazed at the amount of sharing within the ham community.

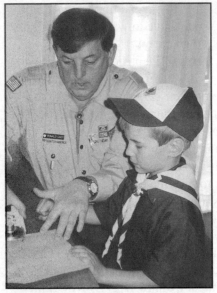

K1RKD assisted numerous Scouts of all ages at the 2004 Jamboree On The Air station. Scouts can earn the Radio merit badge any time!

YOUR RADIO CLUB

Let's start with your local radio club. If you haven't already located a radio club, you can find clubs by using the ARRL's Affiliated Club Search at (**www.arrl.org/clubsearch**). There are several types of clubs; some specialize in one type of operating or public service. Most are "General Interest" clubs for the members to socialize, learn, and help each other out—a good first choice. Don't hesitate to make use of the contact info and attend a meeting! Many clubs make an extra effort to offer special assistance to aspiring and new hams.

Once you decide on a club, you'll get a lot more than just study help by participating in the club activities. Log on to the club Web site. Take advantage of open houses, work parties, operating events, and maybe those informal lunches or breakfasts. Be sure to introduce yourself to the club officers and let them know you're a visitor or new member. Is there another new member? Buddy up! Soon you'll be one of the regulars.

ARRL—THE NATIONAL ASSOCIATION FOR AMATEUR RADIO

Possibly the oldest radio organization in the world, the American Radio Relay League has been an integral part of Amateur Radio from the very beginning. The ARRL offers more assistance to potential and licensed hams than any other organization, which includes operating the largest of the Volunteer Examiner Coordinator and working on behalf of all hams with the FCC and Congress. Members also have access to the extensive ARRL Web site which is home to the ARRL Technical Information Service and numerous operating and educational resources. Membership does not require a license and costs about the same as a couple of large pizzas, with considerably longer-lasting benefits!

What's an Elmer?

A ham radio "Elmer" is a person that personally guides and tutors a new ham through the learning process, both before and after getting a license. It doesn't refer to anyone in particular, just the more experienced hams that lend a helping hand to newcomers. Just about everyone has an Elmer at one point and sometimes several! It's one of ham radio's highest compliments to be someone's Elmer.

The core missions of the ARRL are:

- **Public Service**

The ARRL actively promotes the public-service aspects of Amateur Radio, a tradition that has earned respect through decades of service. The ARRL's legacy of public service began in 1935 with the creation of the Amateur Radio Emergency Service, better known as *ARES*, to provide communication support during natural and man-made disasters.

- **Advocacy**

The ARRL represents Amateur Radio in the halls of power at the local, state and federal levels. Thanks to the efforts of the ARRL, Amateur Radio has been able to thrive despite repeated attempts to restrict its growth. The ARRL serves as a voice for Amateur Radio before influential groups such as the Federal Communications Commission and the World Administrative Radio Conference.

- **Education**

The educational mission of the ARRL is twofold. (1) To recruit new amateurs, the ARRL publishes books and study guides for Amateur Radio license exams, maintains a mentor program for new hams and much more. (2) The ARRL also promotes ham radio in school classrooms, advocating its use as a tool to teach science and technology. To that end, the ARRL assists teachers with appropriate instructional materials and training.

- **Membership**

The majority of active amateurs belong to the ARRL, and for good reason! In addition to all the member benefits listed above, ARRL members receive *QST* magazine each month. *QST* is the authoritative source for news and information on any topic that's part of, or relates to, Amateur Radio. In each colorful issue you'll find informative Product Reviews of the newest radios and accessories from hand-held and mobile FM radios, to home-station transceivers, antennas and even shortwave radios. Each month's Coming Conventions and Hamfest Calendar columns show you who's getting together at hamfests, conventions and swapmeets in your area. Whether you're interested in contesting, DXing, or radios, accessories and antennas you can build at home, *QST* covers them all: New trends and the latest technology, fiction, humor, news, club activities, rules and regulations, special events and much more.

Glossary

These words are discussed more thoroughly in the next section, but are defined here to help you

Frequency—the rate at which a radio signal oscillates (electrical vibrations)

Band—a range of frequencies, usually used for a common purpose, such as by amateurs or broadcasters

Signal—radio energy used in a circuit or as a wave in order to communicate

Channel—a number that refers to radio operation on a fixed frequency. Channel numbers are easier to remember than frequencies.

HF, VHF, and UHF—High Frequency, Very High Frequency, and Ultra High Frequency refer to different frequency bands.

Who Made the "Ham"?

How did "amateur" become "ham"? The real answer is unknown! Even before radio, telegraphers referred to a poor operator as a ham. Perhaps this was derived from a poor operator being "ham-fisted" on the telegraph key—an operator's "fist" referred to his or her distinctive style over the wires. With all radio stations sharing the same radio spectrum in the early days, commercial and military operators would sometimes refer to amateurs as hams when there was interference. Regardless, amateurs adopted the term as a badge of honor and proudly refer to each other as "hams" today.

Amateur Radio or "ham" radio will surprise you with all its different activities. If you've encountered Amateur Radio in a public service role or if someone you know has a radio in their home or car then you already have some ideas about ham radio. Maybe you have seen ham radio in a movie screen or read about it in a book. Whatever your experiences with ham radio, you'll find them and many others on the airwaves every day. Amateur Radio is the most powerful communications service available to the private citizen anywhere on Earth—or even above it!

Amateur Radio is a recognized national asset, providing trained operators, technical specialists, and emergency communications in time of need. It was created to give a home to people just like you that have an interest in radio. Some prefer to focus on the technology and science of radio. Competitive events and award programs attract others. Some train to use radio in support of emergency relief efforts or to keep in touch with family. There are hams that just like to talk on the radio, too! This introductory section of *The ARRL Ham Radio License Manual* will give you a broad overview of Amateur Radio so you can understand how radio works and why hams do what they do. Let's start at the beginning, shall we?

BEGINNINGS OF HAM RADIO

Amateur Radio has been around since the beginning of radio communications. It wasn't long after Marconi spanned the Atlantic in 1901 before curious folks began experimenting with "wireless." Amateur Radio more or less invented itself, right along with broadcasting and wireless telegraphy. The very first amateur licenses were granted back in 1912 and the number of "hams" grew rapidly. Early stations used "spark", literally a vigorous and noisy electrical arc, to generate radio waves. Inefficient and hazardous, spark was soon replaced by far more effective vacuum tube transmitters. By the end of the 1920s both voice and Morse code could be heard on the airwaves. Radio took the world by storm, instantly connecting communities and individuals as they had never been before.

As radio communication became widespread, the Federal Communications Commission was created to regulate the competing radio users; broadcasters, commercial message and news services, military, and public safety. The Amateur Service (the legal name for Amateur Radio) was created in 1934 and has expanded in size and capability ever since. Amateurs, skilled in the ways of radio, played crucial roles during World War II as operators and radio engineers.

After the war, thousands of hams turned to radio and electronics as a profession, fueling the rapid advances in communications during the 1950s and 60s. Amateur radio evolved right along with industry—spanning the globe was commonplace! With Morse code as popular as ever, the amateur frequency bands were also filling with voice and radioteletype signals. Hams even invented a new form of picture transmission called

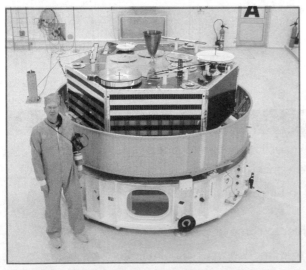

Hams have been building "OSCAR" satellites for decades. OSCAR stands for Orbiting Satellite Carrying Amateur Radio. The first OSCAR was launched in 1961!

KE7AYY works into the night filling the 20-meter log during the W7VMI Field Day at 2004. (AH6B Photo)

AB2LE is studying computer science at college. He earned his license when he was 12.

First licensed at 7 years old, freshman engineering student KBØVVT loves CW contesting.

Operating from a tent on the beach, WA1S made over 5000 contacts from isolated Kure Atoll in the Pacific with the K7C team in 2005. (WA1S Photo)

slow-scan television that could be performed with regular voice equipment. The first amateur satellite, OSCAR-1, was launched in 1961, transmitting a simple Morse message back to Earth for several weeks.

Through the 1970s amateurs built an extensive network of repeater stations, providing regional communications with low-power mobile and handheld radios. In the 1980s and 1990s, microprocessors were quickly applied to radio, greatly increasing the capabilities of amateur equipment and ushering in a new era of digital communications. Packet radio, an adaptation of computer networks, was developed by hams and is now widely used for commercial and public safety. The personal computer, as in many other fields, made huge inroads to ham radio giving amateurs a powerful new tool for design, modeling, and recordkeeping, as well as making amateur radio computer networks a reality. Finally, the Internet arrived and hams quickly adapted the new technology to their own uses just as they had many times before. At each step in the development of today's communication-intensive world, hams have contributed either as part of their profession or as individuals pursuing a personal passion.

HAM RADIO TODAY

Here we are a century later and wireless is still very much at the forefront of communications technology. Far from being eclipsed by the Internet, ham radio continues its tradition of innovation by combining the Internet with radio technologies in new ways. Hams have created their own wireless data networks, position reporting systems, and even a radio-based email network that enables the most solitary sailor to "log in" from anywhere in the world. Voice communications hop between Internet and radio links to connect hams on the opposite sides of the globe using only handheld transceivers less powerful than a flashlight!

Don't let anyone tell you that Morse code is finished! It's still very much alive on Amateur Radio where its simplicity and efficiency continue to make it popular. Amateurs also speak to each other directly using sophisticated radios that are grown-up versions of the Citizen's Band and Family Radio Service radios available at the local electronics stores. Computers are a big part of ham radio today, too, as hams chat "keyboard-to-keyboard" or send pictures via radio. You'll even find some hams assembling their own TV stations and transmitting professional-quality video!

In step with the telecommunications industry, hams also look to the skies for their communications. There are more than a dozen active amateur radio satellites whirling through orbit, connecting hams on the ground by voice, Morse code, and data signals. There is even a ham station on the International Space Station used by astronauts (most are hams) and ground-based hams alike. Ham-written software allows signals to be bounced off the moon and even meteor trails in the Earth's atmosphere.

When disaster strikes, you find hams responding quickly and capably in support of public safety agencies and relief organizations such as the Red Cross. Amateur Radio is an important part of of many emergency communications plans. Between emergencies, hams turn out in great numbers to provide communications for parades, sporting events, festivals, and other public occasions.

Welcome to Amateur Radio

WA0RJY loves the Sweep-stakes contest and operating mobile around Washington state. (N0AX photo)

While Amateur Radio got its start "way back when" as a collection of tinkerers in basements and backyard "shacks", it has grown to become a worldwide communications service for millions of licensees. The tinkerers are very much still with us, of course, creating new and useful ways of putting radio to work. You will find that ham radio has more aspects than you could imagine and they're growing in number every day. Are you ready to join us?

WHO CAN BE A HAM?

Anyone can become a ham. It doesn't matter how old you are or how much you know about radio when you begin. One of ham radio's most enjoyable aspects is that you're on a first-name basis with every other ham, whether you're an elementary school student, a CEO, an astronaut, or a long-distance truck driver. There are also thousands of people with disabilities for whom ham radio is a new window to the world.

Hams range in age from six to more than one hundred years old. While some are technically skilled, holding positions as scientists, engineers, or technicians, all walks of life are represented on the airwaves. Musician? Try Patti Loveless, KD4WUJ, country music superstar, or Joe Walsh, WB6ACU, guitarist for the Eagles. Nobel prize winner? Meet Joe Taylor, K1JT. Athlete? Now playing, Joe Rudi, NK7U, retired major league outfielder and MVP! Why not imagine your name and call sign up there in lights, too?

Hams in Space—Astronauts on the Air

Can you be on the air if you're travelling many miles above the air? Yes and the NASA and Russian astronauts will attest to the value of ham radio while looking down on the Earth from orbit. Nearly every one of the current astronauts has passed at least one level of amateur licensing. Both the Space Shuttle and International Space Station have amateur equipment aboard and if their schedule allows it, the astronauts will make contacts with ground-dwelling hams. The popular ARISS, or Amateur Radio-ISS program (**www.arrl.org/ARISS**) puts schools and other public groups directly in touch with the astronauts via Amateur Radio. Contacting the astronauts takes no more than a handheld radio and some simple antennas!

HANDI-HAMS—No Barriers to Ham Radio

Amateur Radio presents an opportunity to communicate and participate for those with disabilities. Disabled hams often use their unique talents to make valuable contributions over the airwaves, even though they might not be physically present or active. The Courage Center, of Golden Valley, Minnesota, sponsors the HANDI-HAM system (**www.handiham.org**) to help people with physical disabilities obtain amateur licenses. The system provides materials and instruction to persons with disabilities interested in obtaining ham licenses. The Center also provides information to other hams, dubbed "verticals," who wish to help people with diabilities earn a license.

When he's not practicing drums, doing homework, or hosting an Internet radio show with his dad, Drew Pine, KE7AYY, finds time to operate on 2 meters and 70 cm. Drew was first licensed at the age of 11 and immediately joined his local club. Drew likes to operate on Field Day and also supports public service events, such as during a triathlon, where he handled backup net control duties.

Tim O'Donnell, AB2LE was first licensed at age 12 after being introduced to ham radio by a neighbor. Now an Amateur Extra, at age 18 Tim was the ARRL's Goldfarb Scholarship winner for 2005 and enrolled in college as a computer science student. Tim likes to operate using voice on the 20-meter band—he built his dipole antenna himself with his dad John, KC2HHY. Tim's got competition in the family; his 12-year-old sister Genie is KC2MRY!

17-year-old Rebecca Rich,

W7PDZ operates on both HF and VHF from home and from his RV when he takes a road trip. (W7PDZ photo)

KB0VVT has been very active on the airwaves ever since getting her license at the tender age of 7. Encouraged by her parents Dave, KG0US, and Barb, KG0UT, Rebecca is often heard operating in contests on weekends and really enjoys Morse code. Rebecca is one of a handful of hams that reached Amateur Extra while younger than 10! Rebecca helps other kids get into ham radio and was the ARRL's Hiram Percy Maxim award winner for 2005.

During the week Ann Santos, WA1S, is a Information Technology Specialist for the Federal Aviation Administration. On the weekends, she's an active DXer (contacting different countries around the world) or entering a contest. Ann just completed a "DXpedition" to Kure Atoll in the Pacific in 2005 with a group that made 50,000 contacts over two weeks! She's also a member of several ham radio women's organizations, such as the Young Ladies Relay League (YLRL).

Jack Fleming, WA0RJY, originally hailed from the state of Iowa, but after some travels settled down in Seattle, Washington. Jack is an Executive Assistant to a top official at one of Seattle's major hospitals and the editor of the Western Washington DX Club's award-winning monthly newsletter, *The Totem Tabloid*. A Seattle Mariners baseball fan, Jack also manages a fantasy baseball week when not operating mobile or low power portable around the state.

Looking for a hobby at the age of 63, Phil Zook, W7PDZ took a licensing class and passed his Technician license exam. The island club members helped Phil get started and soon he joined the ARES, MARS, and SATERN programs, too. Phil is now a General Class licensee and operates on both HF and VHF. He often takes road trips in his RV with "second op" Kirby the Beagle, keeping in touch with ham radio and sometimes helping fellow travelers along the way.

WHAT DO HAMS DO?

There are so many things that hams do that it's almost easier to tell you what they don't do! The image most people have of a ham is someone with headphones hunched over a stack of glowing radios listening to crackling voices from around the globe. We certainly do that—thousands of contacts span the oceans every day—but there is so much more!

Talking

Hams talk, literally, more than any other way of communicating. After all, it's the way humans are built. For that reason, most amateur radios are made for voice communication. From the glowing radios of yore to the miniature hand-held models that fit in a shirt pocket, connect a microphone or use one built-in and you'll be on the air. Hams use relay stations called repeaters to allow hams using low-power radios to talk with each over a wide region, 50 miles or more. On other frequencies, hams bounce signals off the upper layers of the atmosphere, thousands of miles around the globe. Even computer-to-computer speech is available on ham radio; across town or to a distant repeater system on another continent.

K3OOO (left) and Ray, N9JA, (right) operate the ARRL station, W1AW, using voice during 2004's annual Sweepstakes contest.

Sending

The oldest method hams use is the venerable Morse code, now well into its second century of use. Far from being a useless antique, Morse transmissions are simple to generate and very efficient. Hams also love the musicality of "the code" and have devised many ways to generate Morse, from the original "straight key" to fancy electronic versions and even via computer keyboards. Morse has a language all its own, both in the clear, pure tones and in the mannerisms with which hams connect and converse.

K1DJW, W4OZK and K1DAV load their gear at a marshalling center in Alabama, ready to head into the hurricane Katrina response effort.

KM5WN relays requests for supplies from some of the hundreds of EOC offices in the disaster zone.

Byte-ing

Hams also use digital "codes" to communicate. By hooking up a computer's sound card to a radio or using a *data interface*, digital information can be sent over the air waves. This is one of the most exciting facets of 21st-century Amateur Radio. Not only are hams sending digital data, they're inventing new ways of doing it!

Building

Unlike many other types of radio communications, you are allowed—no, encouraged—to build and repair your own equipment, from the radio itself to the antennas and any accessory you can think of. Hams call this build-it-yourself ethic "homebrewing" and are proud to use homebrew equipment. In the early days of radio, there was no choice; if you wanted to use radio, you had to build it yourself. Today, you'll find hams using everything from the latest manufactured models to a miniature, hand-built radio kit that fits in a metal mint tin! Hams love to build antennas, too, and have been responsible for numerous advances in the state of the antenna art as they tinker and test. If you like to know what's "under the hood" you'll find many like-minded friends in ham radio.

Watching

Hams have devised many ways of exchanging pictures and video. Many years ago, hams used teletype to send pictures made up of text characters! Today a ham can set up a video camera and transmit pictures every bit as good as professional broadcast TV. This is called ATV for Amateur Television and there are even ATV transmitters flying in model aircraft! Hams pioneered the use of regular voice transmitters to send still photographs as "slow-scan television" over long-distance paths to hams thousands of miles away.

Emergency Communications & Public Service

One of the reasons Amateur Radio continues to enjoy its privileged position on the airwaves is the legendary ability of hams to organize and respond to disasters and emergencies. "When All Else Fails…" is a popular ham motto and it's true! Because amateur radio doesn't depend on extensive support systems, ham stations are likely to be able to

Picnic, Training, or Contest? It's ARRL Field Day!

The most popular ham radio event of all is called ARRL Field Day, held on the fourth full weekend of June every year. More than 30,000 hams participate across the United States, Canada, and the Carribean. (European clubs hold their own Field Day on different weekends.) The goal is to exercise your abilities to set up and operate radio equipment "in the field" as if an emergency or disaster occurs. Field Day takes on all the aspects of a competition, emergency exercise, and club picnic —it's all three! Visitors, particularly prospective hams like you, are welcome at these events and there's no better way to learn about ham radio. See **www.arrl.org/FieldDay**.

Katrina 2005—Hams Volunteer Across America

Amateur Radio Emergency Service (ARES) volunteers in Louisiana were heavily engaged in the Hurricane Katrina recovery effort, and more were waiting in the wings to help as soon as they were cleared to enter storm-ravaged zones. Winds and flooding from the huge storm wreaked havoc in Louisiana, Mississippi and Alabama after Katrina came ashore early Monday, August 29, 2005. Louisiana ARES Section Emergency Coordinator Gary Stratton, K5GLS, told the ARRL that some 250 ARES members had been working with relief organizations and emergency management agencies from the beginning.

In Alabama, ARRL Section Manager Greg Sarratt, W4OZK, reported power outages in the northern part of the state when Katrina moved through the region Monday evening, bringing flooding rains and high winds. Amateur Radio SKYWARN nets were active Monday, reporting the severe weather conditions to the National Weather Service. Sarratt himself handled a volunteer shift at the Huntsville NWS office Monday evening. He told ARRL Headquarters that ARES groups throughout the state—and especially in central and southern Alabama—had been supporting communication for local emergency management agencies and the Red Cross.

ARRL Mississippi Section Manager Malcolm Keown, W5XX, in Vicksburg, remained on the air using generator power and makeshift antennas. His area was initially without electrical power or telephone service. Assistant Mississippi Section Manager Edwin Franks, AD5IF, became a nerve center for people call to get information about friends and relatives in the stricken area.

South Texas SEC Jerry Reimer, KK5CA, says Amateur Radio operators deployed from Houston for a communications assignment at the New Orleans Superdome. Because of additional flooding, damage to the facility and other problems at the Superdome, authorities relocated some 25,000 flood evacuees from the Superdome by bus convoy to the Houston Astrodome. Hams assisted with that effort, which took two or three days.

These stories offer just a tiny glimpse of how Amateur Radio responded to this monumental disaster. Amateurs came to help from throughout the United States. In the February 2006 issue of *QST* magazine, the membership journal of the ARRL, you find a list of *more than 1000* amateurs who assisted during the that disastrous hurricane season.

operate while the public communications networks are recovering from a hurricane or earthquake. Hams are also self-organized in teams that train to respond quickly and provide communications wherever it's needed. It's not necessary to have a big emergency for hams to pitch in, we're also providing public service by assisting with communications at parades, sporting events, or serving as weather watchers.

Hams recognize the value of Amateur Radio to their communities and have created training programs, such as the ARRL Amateur Radio Emergency Service® (ARES®), that promote readiness. The ARRL offers several Emergency Communications (or "Emcomm") training classes over the Internet, too. Hams often work closely with other citizen volunteer teams, with many also certified as emergency response workers with a wide variety of skills such as first aid, search-and-rescue, and so forth. Hams hone their message-handling skills in "traffic" nets that pass routine messages around the country so that they'll be ready when called upon for real.

WHAT MAKES AMATEUR RADIO DIFFERENT?

There are lots of other types of radios you can buy in a store—Citizen's Band, handheld FRS/GMRS "walkabouts", marine radio for boaters—what is it about Amateur Radio that sets it apart? In a word, variety. You'll find that each of the radio types listed below are designed for just a few purposes and they might do that well. Amateur Radio, on the other hand, is tremendously flexible with many different types of signals and radio

Table 1-1
Types of Personal Radio

Service	Channels	Intended Use	Range
Citizens Band (CB)	40	private/business	10 miles +
Marine VHF	50	maritime	20 miles +
Family Radio Service (FRS)	22	Personal	2 miles
Multi-Use Radio Service (MURS)	5	Personal	5 miles+

That's Why It's Called "Amateur" Radio

In order to keep businesses or municipalities from unfairly exploiting the amateur bands, amateurs are strictly forbidden from receiving compensation for their activities. That means you can't talk with a co-worker about an assignment, for example. If you provide communications for a parade or charity activity, you can't accept a fee. Your employer is not allowed to pay you for time spent using an emergency ham station in a disaster-preparedness drill. This keeps Amateur Radio free to explore and improve and train. It's worked well for many years!

bands. As a ham, you can pick the right combination of radio signal and band that suits your purpose. You're not restricted to any one combination; you can experiment and try different things as much as you want.

Unlicensed Personal Radios

The most popular personal radios are the FRS/GMRS handheld radios that are seemingly sold everywhere. FRS stands for Family Radio Service and GMRS stands for General Mobile Radio Service. These radios use a set of 22 channels in a narrow frequency band best suited for short-range, line-of-sight communications. (You may be unaware that using the GMRS channels and features of the radio requires a license! It's in the manual in fine print.) Without the GMRS license, your maximum ½-watt of transmitter output power limits you to communications over a few hundred yards to a couple of miles. You can't extend your range with repeaters, nor can you use more powerful mobile radios.

Although the big "Good Buddy" fad of the 1970s is history, Citizens Band (CB) remains popular in the applications for which it was originally intended. Mobile radios in vehicles, boats, and farm equipment provide useful, medium-range radio communications to base radios at home or at work. Handheld radios are also popular. CB radios have 40 channels (more with selectable sidebands) in a frequency band in which communications is fairly reliable over a range of several miles.

Boaters will be familiar with marine VHF radios used for boats to communicate with each other and with stations on shore. There are 50 channels in a frequency range close to the amateur "two-meter" band and mixed in with a public safety and business band. These radios are useful around harbors and for both fresh and salt-water travel. Power restrictions and the choice of frequencies keep these radios in the short-range category

All three of these radios are designed to use a set of channels selected for a single type of communications as shown in **Table 1-1**. They do their designated job well. Amateurs have access to a much broader range of frequencies however, including some near each of these radio services. In many instances, hams use their radios in similar ways to those of CB, FRS/GMRS, and marine. Hams also create new ways of communicating that are more powerful and flexible than those of the unlicensed radio users. If you find your personal radio interesting, but limited, then Amateur Radio is definitely the place for you.

Business and Public Safety Radio

Every day you see police and firemen using handheld and mobile radios as part of their jobs. Many businesses also have their employees use similar radios. How do these relate to Amateur Radio? The FCC has created four large frequency bands with hundreds of available channels for the use of public agencies and private businesses. These users, public and private, are all licensed, just like hams must be.

As with the unlicensed services, amateurs use nearby frequencies that have similar characteristics. The radio electronics are very similar to amateur radios, sometimes identical. In fact, many a ham has adapted a surplus commercial or public safety radio to ham radio use. The commercial "model" of radio use based on discrete channels has been adopted by hams in portions of their nearby bands. However, in other portions of those bands, hams also use types of signals and methods that are completely unknown to their channelized cousins.

Before you go on, study test question T1D01.

1.4 THE FCC & Licensed Radio Services

The Federal Communications Commission (FCC) is charged with administering all of the radio frequencies used by U.S. radio stations. The FCC also coordinates the use of these frequencies with other countries as part of the International Telecommunications Union (ITU). While you may have used an FRS or CB radio without a license, the vast majority of radio users must have a license or be employed by a company that has a license. This section explains how licensing works for Amateur Radio.

If you can learn the basics of radio and the rules of Amateur Radio, then the banquet of ham radio is all yours!

Amateurs have great latitude in how we use our radios.

WHY GET A LICENSE?

Amateurs are free to choose any combination of radio and activity—that's what you get in return for taking the license exam. If you can learn the basics of radio and the rules of Amateur Radio, then the banquet of ham radio is all yours! Just remember that the license is there to insure that you understand the basics before pushing the transmit button. This helps keep Amateur Radio useful and enjoyable to everyone.

Why don't people just buy radios and transmit anyway? (This is called bootlegging or pirating.) Because it's quite apparent to hams who has and who hasn't passed a license exam. You'll find yourself attracting the attention of the Federal Communications Commission (FCC), but more importantly, you won't fit in and you won't have fun.

One of the most important benefits to being licensed is that you have the right to be protected from interference by signals from unlicensed devices, such as consumer electronics. Your right to use the amateur bands is similarly protected. The protection doesn't work perfectly all the time, but nevertheless, as a licensed amateur operator, your license is recognized by law. This is a big improvement over unlicensed radio users. It's definitely worth the effort to get that license!

LICENSING OVERVIEW

The FCC has a different sets of rules for each type of radio use. These uses are called *services*. Each service was created for a specific purpose; Land Mobile, Aviation, Broadcasting. The majority of all services require that a license be obtained before transmissions can be made. These are called *licensed services* and the Amateur Service is one such service.

Most services do not require an examination to be licensed. This is because the FCC sets strict technical standards for the radio equipment used in these services and restricts how it may be used. This tradeoff lowers the training required for radio users. Licensing in these services is primarily a method to control access to the airwaves.

Amateurs, on the other hand, have great latitude in how we use their radios. We can build and

The Upgrade Trail

Successfully obtaining your Technician license is a great achievement—enjoy it! After you get some experience on the air and interact with your fellow hams, it's likely that you'll become interested in operating on the HF bands. The privileges earned by General and Amateur Extra ticket holders enable them to make contacts over really long distances on a wide variety of frequencies. This is where digital mode experimentation is the most active, too. It's a whole different experience on the "short waves." Fortunately, there are a lot of resources available as you hit the "upgrade trail." The ARRL offers study guides and local clubs often sponsor classes. The same online test sites feature General and Amateur Extra tests.

The important thing is to just get started and keep the wheels turning. Read some of the books listed as resources. Keep a study guide handy. Ask questions and visit stations where you can use HF. Be sure to attend a Field Day event where all of ham radio is often on display. Just like studying for your Technician license, soon it starts making sense and your scores on the practice exams will soar.

repair our own radios. The procedures on the air are completely up to us. We can operate wherever we want in our bands, with few restrictions. This flexbility, in order to not cause interference to other radio services, requires that the radio users be more knowledgeable than the typical user in other services. Amaters, therefore, have to pass a licensing examination.

Amateur Licenses

Once upon a time, the FCC gave Amateur Radio exams. You'll hear tales of how hams had to travel long distances to get to the regional Federal Building, stand in lines for hours, sit on uncomfortable chairs, and sweat their way through the exams which were graded by grim-faced proctors. It's a wonder any of us survived!

Today, amateurs give and grade the exams. We even get to make up our own license questions as a team called the National Council of Volunteer Examination Coordinators (NCVEC)! That doesn't mean you won't sweat a little bit when you're taking the exam, but the licensing process is not as imposing as it seems. The result is an *operator license* for each ham. The license (or *ticket*) is *granted* for a term of 10 years after the FCC receives notice from a Volunteer Examination Coordinator (VEC) that you have passed the necessary *elements*. (Elements refers to the three levels of licensing tests, plus the Morse code exam.) The license also specifies your call sign that becomes your radio identity.

There are three classes of license being granted today; the Technician, General, and Amateur Extra. Each grants the licensee more and more *privileges*, meaning frequencies and modes that can be used to make contacts. **Table 1-2** shows the elements and privileges for each of the Amateur license classes as of early 2006. We'll cover those privileges in detail later, but for now all you need to remember is that Technicians are restricted to VHF and UHF bands, while General and Amateur Extra licensees have access to the traditional "short wave" or HF bands.

As a licensed operator in the Amateur Service, you are expected to comply with all of the FCC's rules for the Amateur Service. In return, you are granted the right to operate and to be free from harmful interference by unlicensed users and devices.

Table 1-2
Amateur License Class Examinations

License Class	Test Element Designation	Number of Questions	Privileges
Technician	2	35 (passing grade is 26 correct)	All VHF and UHF privileges
General	3	35 (passing grade is 26 correct)	All VHF, UHF and most HF privileges
Amateur Extra	4	50 (passing grade is 37 correct)	All amateur privileges

1.5 Basic Amateur Radio Activities

Ham radio has a lot to offer but the many activities can be confusing. To help you understand why we operate in certain ways or why rules are written the way they are, this section presents some basic ham radio. Later, you'll learn more details, but this introduction shows you some of the fundamentals that are present in almost all ham radio communications.

IDENTIFICATION AND CONTACTS

On the airwaves, your everyday identity gets something new—a *call sign*. Instead of "Bob" or "Mary", your primary radio identity becomes "Steve WB8IMY" or "Mary K1MMH." Hams become known by their call signs and often keep them for life. Your call sign or *call* is completely unique among all the radio users anywhere in the world! There is only one N6BV (Dean in USA) just like there is only one G4BUO (Dave in England) and one I2UIY (Paolo in Italy) and one JE1CKA (Tack in Japan). By transmitting your call, other hams know who you are and your nationality. Identifying yourself with your call sign is known as *signing*. Because you can't see other hams except when using video transmissions, your call sign is very important. In fact, you're required to state your call regularly during every contact so that everyone knows whose transmissions are whose.

Speaking of contacts, any conversation between hams over the air is called a *contact* and starting a conversation is *making contact*. Attempting to make contact by transmitting your call sign is *making a call* or *calling*. If you're making a "come in anybody" call to which any station can respond, that's *calling CQ*. ("CQ" means "a general call.") If you've made a contact with someone, you've *worked* them. Hams also use a radio shorthand, called "Q-Signals," in which a contact becomes a "Q-S-O." (You'll meet more Q-Signals later.)

A frequent question is "How do you know where to get in touch with another ham?" Hams are generally not as confined in their frequency choices as other radio users. On the bands where repeaters operate, there are known channels, but you still have to know which channel the person that you're trying to contact is operating. There are lots of channels—more than on any commercial radio! On other bands, there are no channels at all and hams are free to tune their radios to wherever they can make contact. It sounds like chaos, but hams have worked out methods of keeping things orderly.

When calling CQ, you literally say "CQ" as in, "CQ CQ CQ, this is W1AW calling CQ." If you hear a station calling CQ, you can make a contact with them by transmitting their call and then saying "this is (your call)," such as "W1AW this is K5TR." Once you establish contact, the next step is to exchange more information such as a *signal report* that lets the other station know how well you're receiving or *copying* them. Name and location are exchanged after that—then you're off to whatever business is at hand. A long, friendly conversation is known as a *ragchew*. At the end of a contact, you "sign off"

Ham Shorthand

Like any activity that has been around for a while, such as sailing or flying, radio has a special jargon all its own. Many of these conventions originate from the days of the telegraph. In those days, every word took up precious time, so the operators developed an extensive series of abbreviations and special characters (called *prosigns*) that kept the information (or *traffic*) flowing quickly and smoothly. For example, you may have heard the word "break" or "breaker" used over the radio. The word originally referred to a telegraph operator disconnecting or *breaking* the telegraph line so that no characters could be sent. This got the attention of every other operator along the line and to do so was *breaking in*. The word is still in use 150 years later!

Later, as radio became a worldwide tool, operators that didn't speak the same language were aided by the creation of Q-signals. For example, "QTH?" means "What is your location?" and in response "QTH Seattle" means "I am located in Seattle." Many of these procedures and abbreviations are still in place today, because they work!

although it's mostly movie imagination that radio operators say, "signing off."

Why so much procedure? It's important to remember that radio communications, unlike using the telephone, are one-at-a-time conversations. If I'm transmitting, I can't hear you. Only one person can transmit at a time. Neither can you be seen by the other hams, so all of the visual clues from our faces and hands that we use in daily life are missing. All that we know of the operator on the "other end" comes out of a speaker or headphones. This requires "radio etiquette" so that we can share the airwaves efficiently. Speaking clearly (remember, it all has to be understood through that little speaker) and frequently identifying your station are the first two steps. Next you learn to use procedural signals such as *over*, *break*, and *clear* correctly. With practice, you'll become comfortable using radio jargon, such as that in the glossary at the back of the book. As you learn about the various ways that hams communicate, remember that those contacts all share the same fundamental etiquette.

USING YOUR VOICE

By far, the most popular method or *mode* of making contacts is by voice. There are a number of ways to transmit voices via radio signals and you'll meet those in Part II. It's easy and natural to converse this way, using the proper procedures. Voice is widely used by amateurs for short-distance and long-distance contacts. It is far and away the most popular mode for hams on the go and during public service or emergency response.

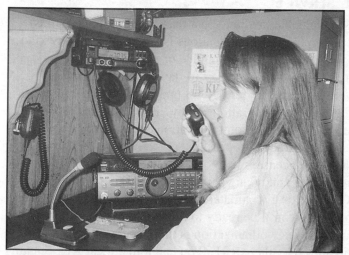

Hams use repeaters to relay signals from low-power radios over a wide area. Repeaters are a popular on-the-air meeting place for hams.

As a Technician licensee, you'll be able to make voice contacts directly with other hams and also by using stations called *repeaters* that relay signals from low-power mobile or hand-held transmitters across a wide area. Hams have also built Internet-linked repeater systems that send digitized voices around the world so that local repeater users can communicate world-wide with just a low-power, hand-held radio! Three examples are IRLP (**www.irlp.net**), Echolink (**www.echolink.org**), and D-STAR (**www.icomamerica.com/amateur/dstar**).

Hams use English as a common language around the world, too, although when communicating within a country hams use their native languages. Voice contacts can be a good way to learn or polish foreign language skills, too.

USING MORSE CODE

Morse is still quite popular in Amateur Radio. Because all of a signal's energy is concentrated in the simple on-and-off signal, Morse works very well in the presence of interference or when signal levels are weak. Morse signals can be generated by extremely simple transmitters—something to generate a radio signal and something else to turn it on and off! Receiving and decoding Morse (called *copying the code*) requires only a basic receiver and a human ear.

Many operators enjoy the rhythm and musicality of "the code," as well. Aside from its utility as a communications protocol, it's a skill like whistling or painting that you can enjoy for its own sake. Listening to a skilled Morse operator chatting away or relaying messages is quite a treat!

EXCHANGING DIGITAL DATA

The recent availability of inexpensive sound-card hardware and signal processing software for personal computers has brought about a surge of interest in the *digital modes* where the conversation is carried out as streams of characters sent over the airwaves. A *data interface* is used to connect the radio to the computer. Most digital contacts are *keyboard-to-keyboard*, meaning that the operators take turns typing, just as in an Internet chat room or on messaging system.

Hams have been blazing trails in developing new methods of converting the computer characters into radio signals and back again. The methods are called *protocols* and are referred to by their initials, such as RTTY, PSK31, AMTOR, PACTOR, or MFSK to name just a few. Different protocols are applied to different types of radio communication because of the effects that transmission and reception have on the radio signals.

One of the new digital radio systems developed by hams exchanges e-mail over Amateur Radio. It's called Winlink and it looks a lot like regular e-mail on the computer screen. It's used daily by thousands of hams that are unable to access the Internet while travelling, at sea, or camping in remote locations.

EMERGENCIES AND PUBLIC SERVICE

One of the reasons ham radio is so valuable (and maybe a reason you're reading this book) is that hams are good at helping. Communications is key to making any kind of organized effort work whether its a small parade or major emergency response to a natural disaster. While the day-to-day telecommunication systems are recovering, hams can quickly set up networks that support public safety and government operations. Why? Because there are lots of hams and we are skilled in basic communications that doesn't depend on lots of sophisticated equipment to operate.

Can Technician Class licensees help out? You bet they can! By learning how to use your radio and taking some simple training classes, such as the ARRL's EC-001 " Level 1 Amateur Radio Emergency Communications" (**www.arrl.org/cce**), you'll be ready to join and practice with other hams. The largest ham emergency organization is the Amateur Radio Emergency Service (ARES), organized by the ARRL. You can join a local ARES team to receive training and practice providing emergency communications support.

Table 1-3 lists several ham radio emergency response groups.

Hams can pitch in and help in many ways. Not everyone has to be on-site to make a contribution. Whatever your personal capabilities and license class, there is a need you can fill.

Table 1-3
Emergency Response Organizations

ARES: **www.arrl.org/ares**	Amateur Radio Emergency Service—organized by the ARRL
RACES: **www.races.net**	Radio Amateur Civil Emergency Service (works with civil defense agencies)
SATERN: **www.satern.org**	Salvation Army Team Emergency Radio Network
HWN: **www.hwn.org**	Hurricane Watch Net—works with the National Hurricane Center
SKYWARN: **www.skywarn.org**	Severe weather watch and reporting system—works with the National Weather Service

- From home—Use powerful base station radios and antennas to provide long-distance communications, relay messages, and act as a *net control* to coordinate communications.
- From a vehicle—From a personal car or a communications-equipped van, portable stations provide valuable relay and net control functions in the field.
- On foot—Go where the action is to provide status reports and relay supply and operations messages between the control centers and workers in the field.

One of the most important functions, repeated three times in the list above, is to *relay* communications. It is no coincidence that the third letter in ARRL stands for "relay," one of the oldest and most highly valued radio functions. Relaying information requires accuracy and efficiency. Hams that provide emergency communications pride themselves on both.

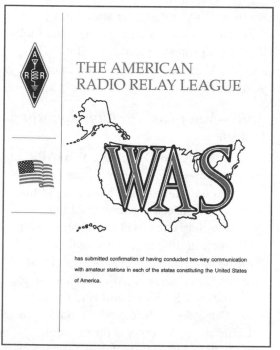

Worked All States (WAS) is often the first major award pursued by hams around the world.

AWARDS AND CONTESTS

Hams keep their skills sharp not only by training, but by getting on the air and having fun! Just as sports and recreational activities keep your body in good health and fit, there are competitive radio activities, too! There are many suitable for Technician license holders, as well.

Awards programs are competitions with yourself. There are operating achievement awards for almost anything you can imagine, such as working (contacting) every state or different countries, contacting satellites, and making low-power contacts. **Table 1-4** shows several examples. For a more complete listing, Ted K1BV maintains a list of more than 3300 awards from around the world in his directory (**www.dxawards.com/book.html**). Collecting these colorful certificates and other prizes can be addictive!

Contests, sometimes called *radiosport*, fill a band with rapid-fire contacts as amateurs strive to make as many short contacts as possible within a limited time. There are a tremendous variety of contests, from short *sprint* events lasting only a few hours to international contests that last for 48 hours! Working hard to compete and improve provides a strong incentive to become skilled in both technical and operating capabilities. Technicians should

---

Table 1-4

Awards & Operating Events for Technician Licensees

OSCAR Satellite Communications Achievement Award:
Contact 20 different states, Canadian provinces, or countries using amateur satellites (**www.amsat.org**)

VHF/UHF Century Club (VUCC):
Contact one hundred grid squares using VHF and UHF frequencies (**www.arrl.org/awards/vucc**)

CQ iDX:
Make contact with DX amateurs using IRLP and Echolink Internet systems (**www.cq-amateur-radio.com**)

ARRL & CQ VHF Contests:
Several contests that make use of the VHF and UHF bands (**www.arrl.org/contest**)

ARRL Field Day:
The largest on-the-air event in ham radio (**www.arrl.org/fieldday**)

definitely check out the ARRL and CQ VHF contests (see the *Ham Radio License Manual* Web page).

NOVEL ACTIVITIES

As if all this wasn't interesting enough, hams are famous for pushing the envelope and either inventing an entirely new technology or adopting a commercial technology in an unexpected way. It was mentioned before that hams have their own satellites and radio-Internet networks, but that's just the start. Here are some examples:

SSTV and ATV

Slow-scan television (SSTV) was invented by hams to send pictures over regular voice radios. Each picture takes about 8 seconds and can be sent and received by a computer with a sound card. Broadcast TV-style video is the domain of the Amateur TV (ATV) enthusiasts. They hook up a regular video camera to an ATV transmitter and, *voila!*, they're on the air, beaming video that looks just like a professional signal. Both ATV and SSTV are increasingly used in emergencies, too.

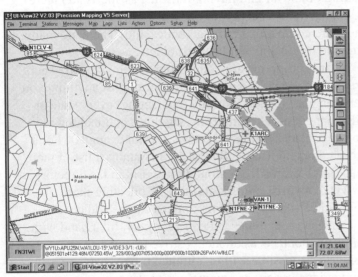

The Automatic Position Reporting System (APRS) was invented by hams to show location and travel path by using ham radio.

APRS—Automatic Position Reporting System

Invented by Bob Bruzinga, WB4APR, as a way of integrating GPS position information with ham radio, APRS was the foundation for the LoJack car theft tracking system widely used today. Amateurs with a GPS and a mobile radio can send their position to a local APRS relay point and Internet-based servers. APRS users can log on to the APRS servers and find the location of anyone sending position data tracking their movements on maps of various detail levels. It's fascinating!

Packet Radio

As computer networks became common, amateurs immediately adapted the popular X.25 computer-to-computer protocol to over-the-air operation. A special kind of data interface called a *Terminal Node Controller* (TNC) takes characters from a computer and re-packages them into data packets which are transmitted by a regular, unmodified amateur VHF radio. Packet networks exist in many cities for hams to share information and some are linked to other packet and computer systems around the world.

Meteor Scatter and EME

Perhaps the most exotic of all ham activities is making contacts via *meteor scatter* or maybe via *Earth-Moon-Earth* (EME) reflections! The trail of a meteor can reflect radio signals during the few seconds it lasts. A skilled amateur can use that trail to make short contacts! The biggest reflector in the sky is the moon. Hams have learned that just like NASA and ESA, they can bounce their signals off the moon and hear them when they complete the round trip back to earth! Nobel Prize-winner Joe Taylor K1JT wrote software that uses a computer's sound card to both send and receive data in a highly specialized code that enables even modest stations to operate both meteor scatter and EME. Does that sound exciting? It is!

Chapter 2

Radio and Electronics Fundamentals

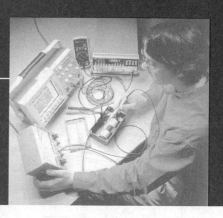

What you'll learn in this chapter:

- **The names and types of basic radio equipment**

- **Fundamental concepts behind electricity and electronics**

- **Common electronic components**

- **What radio signals are and their different forms**

- **Basic concepts of antennas and feed lines**

- **How radio signals travel from place to place**

This is the real beginning of your Amateur Radio Adventure! In this section, we dive into what makes radio work—the equipment, electronics, antennas and the signals themselves. The material in each chapter is presented in a "here's what you need to know" style. References will be provided so that you can learn more about topics that interest you. Start by bookmarking the ARRL's *Ham Radio License Manual* Web page, **www.arrl.org/hrlm**. Don't forget that there's a comprehensive glossary in the back of the book for unfamiliar terms.

Covering these topics first makes it easier for you to understand the material later on. You'll also be a better and safer operator. Relax—we'll start at the beginning and learn one step at a time!

2.1 Equipment Definitions

This chapter covers the different types of basic radio equipment. You've used radios before, of course—at home, in the car, or at work. That means you're already familiar with some of the topics in this chapter. It's a good idea to review though, no matter what your background, since in Amateur Radio the terms might be used a little differently than what you're used to. We'll cover the operating details later. The goal here is to make sure we're using the same words to mean the same thing!

BASIC STATION ORGANIZATION

All radio equipment is designed to generate or manipulate *radio signals*. A radio signal can be a transmission traveling around the world. It can also be the energy inside equipment before or after its journey from place to place. No matter how I intend to communicate with you—with voice, code, or computer—those are radio signals, usually just referred to as *signals*.

There is also a lot of radio energy bouncing around out there that isn't used for communication; static from the atmosphere, electrical noise from computer equipment and motors, buzzing and humming from the power lines to name just a few. When hams talk about signals, though, they're referring to the electrical energy they use to exchange information, inside or out of a radio.

The three basic elements of a radio station, big or small, are the transmitter, receiver, and antenna as shown in **Figure 2-1**. A *transmitter* (abbreviated XMTR) generates a signal that carries speech, Morse code, or data information. A *receiver* (abbreviated RCVR) recovers the speech, Morse code, or data information from a signal. (Figure 2-1 is a *block diagram* that shows how a system of equipment, such as a radio station, is organized without getting into the complex details of every connection and control.)

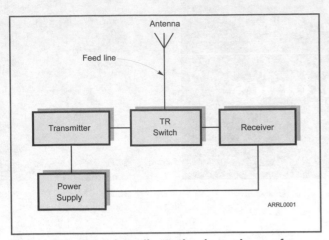

Figure 2-1—A basic radio station is made up of a transmitter and receiver connected to an antenna with a feed line. The transmit-receive (TR) switch allows the transmitter and receiver to share the antenna. A transceiver includes all of those pieces in a single enclosure.

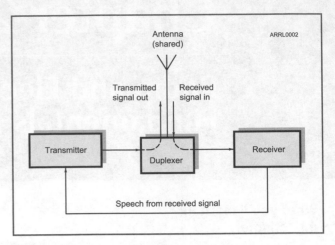

Figure 2-2—In a repeater, audio from the receiver is immediately rebroadcast by the transmitter on a different frequency. The duplexer allows the transmitter and receiver to share the common antenna at the same time.

An *antenna* turns the radio signals from a transmitter into energy that travels through space as a *radio wave*. An antenna also captures radio waves and turns them into signals for a receiver to work with. A *feed line* connects the antenna to the transmitter or receiver. Feed lines are also called *transmission lines*, just like power lines, because they are used to transfer energy—radio signals in this case.

The process of turning the transmitter's output signal into radio waves that leave the antenna is called *radiating*. The radio wave is called *electromagnetic radiation*. Don't let the word "radiation" put you off. Electromagnetic radiation from an antenna is not the same as ionizing radiation from radioactivity, nor does it interact with living organisms in the same way.

Most amateur equipment combines the transmitter and receiver into a single piece of equipment called a *transceiver* (abbreviated XCVR). A transceiver shares a single antenna between the transmitter and receiver circuits by using a *transmit-receive (TR) switch*. If needed, a *power supply* converts external electrical power, such as household ac, to whatever form is needed by the radio equipment.

REPEATERS

Repeaters are a special kind of station that relays signals across a wide area. Repeaters allow low-power handheld and mobile stations to communicate with stations that they can't reach directly. A repeater consists of a receiver and transmitter connected together so that the received signal is re-transmitted as a more powerful signal. Repeaters are located on high buildings, towers, or hills for maximum range. **Figure 2-2** shows the basic elements of a repeater station.

Because a repeater receives and transmits at the same time, instead of a TR-switch it uses a *duplexer*. The duplexer allows the strong signal from the transmitter and the tiny signals the receiver listens for to share a single antenna.

ACCESSORY RADIO EQUIPMENT

With the basic equipment accounted for, now add the accessory equipment for the operator (that's us) to generate or understand the information in radio signals. **Figure 2-3**

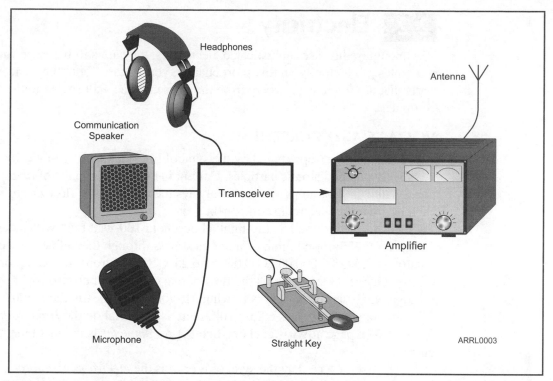

Figure 2-3—Accessory equipment allows operators to send their voices or Morse code and to hear the signals from other stations. An amplifier is used to increase the power of the transmitted signal for better range.

shows the most common accessories used with a basic station.

A *microphone* (or mike, sometimes abbreviated MIC) turns an operator's voice into an electrical signal called *audio*. The transmitter then adds the audio to a radio signal. The microphone is usually built-in on a handheld transceiver but is a separate piece of equipment for larger rigs. A Morse *key* is a special switch used by the operator to turn a transmitter's output signal on and off in the patterns of Morse code.

At the receiver's output, a *speaker* turns the electrical audio signal back into audible sound that the operator can hear. A speaker may be built-in or an external device. *Communications speakers* are specially designed for use with radios to optimize the understandability of speech and Morse code. *Headphones* are often used instead of a speaker to make it easier to understand the audio in the presence of noise or interference and to avoid disturbing others.

Amplifiers are circuits or equipment that increase the strength of a signal. *Preamplifiers* (or preamps) increase the strength of a signal before it is applied to a receiver. *Power amplifiers* (often called linears) increase the strength of a transmitted signal before it is sent to the antenna.

Before you go on, study test questions T4C01-05; T5A01, 02, 04; T5C01. Review this section if you have difficulty.

2.2 Electricity

Although radios use sophisticated electronics, they are still based on fundamental principles of electricity. In this short chapter, you'll learn about the basic electrical concepts that are found everywhere from the household ac wall socket to the latest radio or computer.

VOLTAGE AND CURRENT

Electric *current* (represented by the symbol I or i) is the flow of *electrons*. Electrons are negatively charged atomic particles. Current is measured in units of *amperes* which are abbreviated as A or amps. Current is always measured as the flow *through* something, such as a wire or electronic component.

Electrons are so small that to light a household 100-watt bulb with a current of 1 ampere, 6.25 billion billion of them must pass through the bulb each second! That quantity, 6,250,000,000,000,000,000 or 6.25×10^{18} electrons, makes up one *coulomb* of electric charge. One ampere is the flow of one coulomb of electrons per second (1 C/sec) through whatever the current is flowing. (If you need to learn about numbers written in scientific notation, such as 6.25×10^{18}, there is a tutorial on the *Ham Radio License Manual* Web page at **www.arrl.org/hrlm/**.) An *ammeter* is the instrument used to measure current.

Voltage (represented by the symbol E or e) is the *electro-motive force* or *potential* that makes electrons move and is measured in units of *volts* which are abbreviated as V or v. Voltage is always measured from one point to another or with respect to some reference voltage. Voltage is measured with a *voltmeter*. The *polarity* of voltage can be either positive or negative. A *negative* voltage repels electrons and a *positive* voltage attracts them. The earth's surface acts as a universal reference for voltage measurements and is called *earth ground*, *ground potential* or just *ground*.

Electrical Pressure and Flow

Figure 2-4 shows a time-tested analogy that helps newcomers to electricity understand what voltage and current are and how they act. Voltage is analogous to pressure in a water pipe and current is analogous to water flow. You can have lots of pressure with no flow—think of a pipe with the faucet closed. You can't have flow with no pressure, though. There must always be something pushing water and electrons before they'll move.

Pressure is always measured between two points in the plumbing or between one point and the open air (atmospheric pressure). Voltage is the same—it only has meaning as an electrical force that pushes electrons to flow from one point to another. Water flow, like current, only exists as it flows through something—a pipe, a stream, or a stream from a hose.

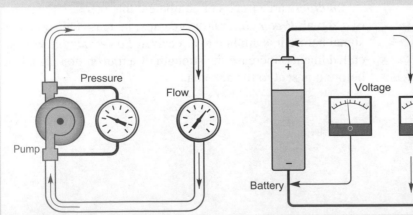

ARRL0004

Figure 2-4—Voltage acts similarly to pressure and current similarly to flow in a water system. Voltage between two points is what causes the electrons to move between those points. Current is a measure of how many electrons pass through the circuit per second.

RESISTANCE

All materials impede the flow of electrons to some degree. This property is called *resistance* (represented by the symbol R) and is measured in ohms which are represented by the symbol Ω, the Greek letter omega. Materials in which electrons flow easily in response to an applied voltage are *conductors*. Metals are good conductors and so is salt water. The human body conducts electricity, too! Materials that resist or prevent the flow of electrons are *insulators*, such as glass and ceramics, dry wood and paper, most plastics, and other non-metals. The *ohmmeter* is used to measure resistance.

Georg Ohm, a scientist that investigated electricity in the 19th century, discovered that voltage, current, and resistance are *proportional*. Ohm's Law states that resistance is the ratio of applied voltage to the resulting current. The more a material resists the flow of electrons, the lower the current will be in response to voltage across the material. As an equation, this is stated R = E / I.

If you know any two of I, E, or R, you can determine the missing quantity as follows:

R = E / I,

I = E / R,

E = I × R.

The drawing in **Figure 2-5** is a convenient aid to remembering Ohm's Law in any of these three forms. **Figure 2-6** shows several examples of how to use Ohm's Law.

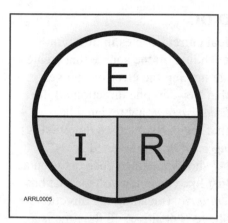

Figure 2-5—This simple diagram will help you remember the Ohm's Law relationships. If you know any two of the quantities, you can find the third by covering up the unknown quantity. The positions of the remaining two symbols show if you have to multiply (side-by-side) or divide (one above the other).

POWER

Power (represented by the symbol P) is measured in watts which are abbreviated as W. Power is the product of voltage and current. As with Ohm's Law, if you know any two of P, E, or I, you can determine the

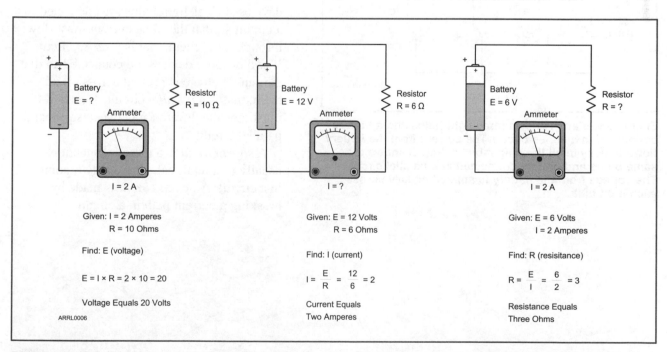

Given: I = 2 Amperes
R = 10 Ohms

Find: E (voltage)

E = I × R = 2 × 10 = 20

Voltage Equals 20 Volts

Given: E = 12 Volts
R = 6 Ohms

Find: I (current)

$I = \dfrac{E}{R} = \dfrac{12}{6} = 2$

Current Equals Two Amperes

Given: E = 6 Volts
I = 2 Amperes

Find: R (resistance)

$R = \dfrac{E}{I} = \dfrac{6}{2} = 3$

Resistance Equals Three Ohms

Figure 2-6—This drawing shows three examples of how to use Ohm's Law to calculate voltage, current, or resistance.

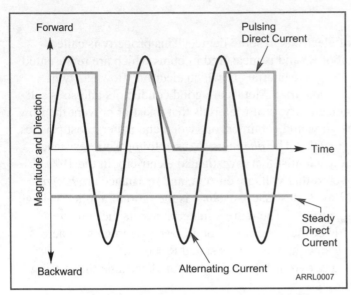

Figure 2-7—Direct current flows steadily in one direction—forward or backwards. Pulsating direct current may stop and start, but always flows in the same direction. Alternating current regularly reverses its direction.

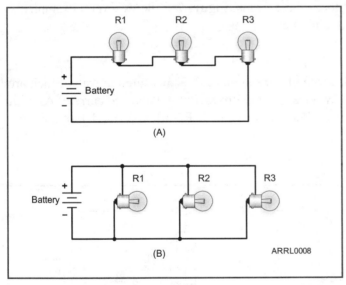

Figure 2-8—Part A shows three light bulbs and a battery connected in a series circuit. The current from the battery flows equally in all three light bulbs. Part B shows the same bulbs and battery connected in a parallel circuit. The voltage from the battery is applied equally across each light bulb.

missing quantity as follows:
$P = E \times I$,
$E = P / I$,
$I = P / E$.
Because Ohm's Law links voltage, current, and resistance, E and I can be replaced in the power equations. Substituting Ohm's Law, $P = I^2 \times R$ and $P = E^2 / R$.

AC AND DC

Electrical current comes in different forms, depending on the source that creates the voltage making the electrons move. Current that flows in one direction all the time is *direct current,* abbreviated dc. Current that regularly reverses direction is a*lternating current*, abbreviated ac. **Figure 2-7** shows the difference between ac and dc. Just like current, a voltage that has the same *polarity* (meaning the direction in which voltage is positive) all the time is a *dc voltage.* A voltage that regularly reverses polarity is an *ac voltage.* Batteries and solar cells are a source of dc voltage and current. Electrical utility power is in the form of ac voltage and current.

CIRCUITS

A *circuit* is any complete path through which current can flow. If two or more devices such as light bulbs are connected in a circuit so that the same current must flow through all of them, that is a *series* circuit. If two or more devices are connected so that the same voltage is present across all of them, that is a *parallel* circuit. **Figure 2-8** illustrates the difference between series and parallel circuits.

A *short circuit* is a direct connection, usually unintentional, between two points in a circuit. An *open circuit* is made by breaking a current path in a circuit.

Before you go on, study test questions T4A01-04; 07-13; T4D01-11; T4E01-06. Review this section if you have difficulty.

2.3 Components & Units

Electronic circuits are made from *components* and the connections between them. Each component performs a discrete function; storing or dissipating energy, routing current, or amplifying a signal. In this chapter you'll learn about the different types of common components and their functions. We begin by reviewing the way components and signals are measured.

METRIC PREFIXES

The units of measurement that describe components and circuit functions use the *metric system* of prefixes. The metric system is used because the numbers involved cover such a wide range of values. **Table 2-1** shows metric prefixes, symbol, and their meaning. The prefixes expand or shrink the units, multiplying them by the factor shown in the table. For example, a kilo-meter (km) is one thousand meters and a milli-meter (mm) is one-thousandth of a meter.

The most common prefixes you'll encounter in radio are pico (p), nano (n), micro (μ), milli (m), centi (c), kilo (k), mega (M), giga (G). It is important to use the proper case for the prefix letter. For example, M means one million and m means one-thousandth. Using the wrong case would make a big difference!

Table 2-1
International System of Units (SI)—Metric Units

Prefix	Symbol	Multiplication Factor
Tera	T	$10^{12} = 1,000,000,000,000$
Giga	G	$10^9 = 1,000,000,000$
Mega	M	$10^6 = 1,000,000$
Kilo	k	$10^3 = 1000$
Hecto	h	$10^2 = 100$
Deca	da	$10^1 = 10$
Deci	d	$10^{-1} = 0.1$
Centi	c	$10^{-2} = 0.01$
Milli	m	$10^{-3} = 0.001$
Micro	μ	$10^{-6} = 0.000001$
Nano	n	$10^{-9} = 0.000000001$
Pico	p	$10^{-12} = 0.000000000001$

If you're already familiar with the metric system, review the questions at the end of the chapter to be sure you have it mastered. If the metric system is unfamiliar to you, the Ham Radio License Manaul Web page has a detailed discussion of how the prefixes work and examples for you to learn from. Review that material until you are comfortable with the examples and definitions.

The Decibel - Bringing the Large and Small Together

Radio signals vary dramatically in size. At the input to a receiver, signals are frequently smaller than one ten-billionth of a watt. When they come out of a transmitter, they're often measured in kilowatts! Electronic circuits change signal strengths by many factors of ten. These big differences in value make it difficult to compare signal sizes. Enter the decibel, abbreviated dB and pronounced "dee-bee." The decibel measures the ratio of two quantities as a power of ten. The formula for computing decibels is:

dB = 10 log (power ratio)

dB = 20 log (voltage ratio).

Positive values of dB mean the ratio is greater than 1 and negative values of dB indicate a ratio of less than 1. For example, if an amplifier turns a 5-watt signal into a 25-watt signal, that's a change of 10 log (25 / 5) = 10 log (5) = 7 dB. On the other hand, if by adjusting a receiver's volume control the audio output signal voltage is reduced from 2 volts to 0.1 volt, that's a change of 20 log (0.1 / 2) = 20 log (0.05) = −26 dB.

A complete discussion of the decibel, its history and tables of easy-to-remember values is available on the Ham Radio License Manual Web page.

COMPONENTS

The three most basic types of electronic components are resistors, capacitors, and inductors (coils). Each have their own units of measurement and have a different effect on voltage and current. The wires that are used to make connections to the component are called *leads* and have only small effects on the performance of the component. In a circuit with more than one component, each has a unique *designator*; a label such as R14, C2, or L20.

Resistors have a certain resistance specified in ohms (Ω), kilo-ohms (kΩ), or mega-ohms (MΩ). The function of a resistor is to oppose the flow of electrical current, just as a valve in a water pipe restricts the flow through the pipe. Resistors, like valves, control flow or current. The electrical current flowing through the resistor loses some of its energy as heat, so resistors also dissipate energy, like an electrical brake. **Figure 2-9** shows different types of resistors.

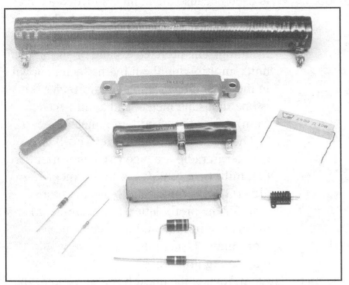

Figure 2-9—This photograph shows some of the many types of resistors. Large, power resistors are at the top of the photo. The small resistors are used in low-power circuits.

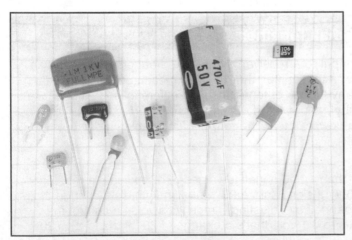

Figure 2-10—This photograph shows a few styles of capacitors that are used in most common electronic equipment. Capacitors used in transmitters and high-power circuits are larger than those shown here.

Capacitors store electric energy. The property of storing electrical energy is called *capacitance*. Capacitors used in radio circuits have values measured in pico-farads (pF), nano-farads (nF), and micro-farads (µF). As it stores and releases energy in a circuit, a capacitor smoothes out voltage changes. Capacitors are made from two conducting surfaces (such as metal foil) separated by an insulator called the *dielectric*. Because the conductors are separated a capacitor cannot pass dc current. AC current, however, can pass through a capacitor. In most capacitors, the metal surfaces and dielectric are sealed inside a protective coating as shown in **Figure 2-10**.

Inductors store magnetic energy. The property of storing magnetic energy is called *inductance*. Inductors have values measured in nano-henries (nH), micro-henries (µH), milli-henries (mH), and henrys (H). As it stores and releases energy in a circuit, an inductor smoothes out current changes. Inductors are made from wire wound in a coil, sometimes around a *core* of magnetic material that concentrates the magnetic energy. For dc voltages, an inductor appears to be a short circuit, but it resists the flow of ac current. **Figure 2-11** shows several common types of inductors.

Many electronic components use paint markings, called a *color code*, to indicate their value. For example, a resistor with four colored stripes, red-violet-brown-gold, was made to have a resistance value of 270 Ω with an accepted variation in that value (called *tolerance*) of 5%. Other components have their value printed directly on their body, but may encode the value to save space. For example, a capacitor labeled with '683' has a value of 68×10^3 pF. You can find out how to read color and marking codes at the link on the Ham Radio License Manual Web

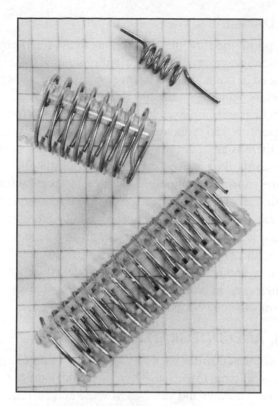

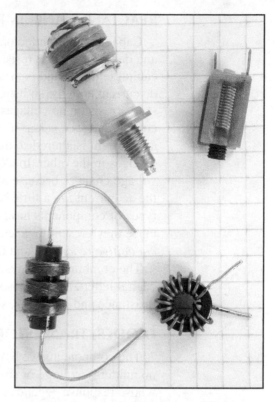

Figure 2-11—Here are two types of inductors. On the left are air-core inductors. Those on the right use a magnetic material to concentrate the stored energy and increase inductance. At the lower right is an inductor wound on a toroid core.

page at **www.arrl.org/hrlm/**.

All three types of basic components are also available as *adjustable* or *variable* models, too. A variable resistor is also called a *potentiometer* or *pot* because it is frequently used to adjust voltage or potential. Variable capacitors and inductors are used to tune radio circuits for a variety of purposes. **Figure 2-12** shows some examples of variable components.

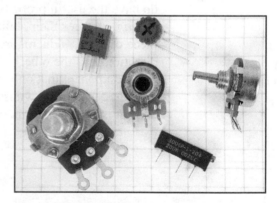

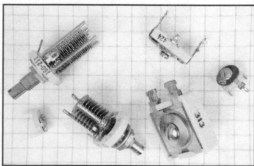

Figure 2-12—Components with adjustable values are used to tune or calibrate circuits. This photograph shows variable resistors (top) and capacitors bottom). Variable inductor can be seen at the top right of Figure 2-11.

Ohm and Farad and Henry - Oh, My!

Electrical units of measure are almost all named for the scientists and experimenters that played an important role in understanding electricity and radio. For example, Georg Ohm (1787-1854) discovered the relationship between current, voltage, and resistance that now bears his name as Ohm's Law and the unit of resistance, the ohm. Table 2-2 lists units and the trailblazers for whom they are named. Complete biographical information on all of these electro-pioneers can be found online in the Wikipedia, **www.wikipedia.org**.

REACTANCE AND IMPEDANCE

Capacitors and inductors oppose the flow of ac current as their stored energy increases and decreases. Their resistance to ac current flow is called *reactance* (represented by the capital letter X) and is measured in ohms, like resistance. A resistor opposes both ac and dc current equally, so its resistance and reactance are the same.

In a resistor, ac voltages and currents are exactly in step: As voltage increases, so does current and vice versa. In capacitors and inductors, the relationship between ac voltage and current is altered so that there is an offset in time between changes in one and changes in the other. In a capacitor, the changes in current are a little ahead of, or *leading*, voltage changes because of the capacitor's smoothing action. In an inductor, changes in the ac current *lag* a little behind changes in voltage for the same reason. Because the two types of components have opposite relationships between voltage and current, the reactance they present to ac current is also different. Reactance from a capacitor is called *capacitive reactance* and from an inductor, *inductive reactance*.

The combination of resistance and reactance is called *impedance* (represented by the capital letter Z) and is also measured in ohms. Radio circuits almost always have both resistance and reactance, so impedance is often used as a general term.

DIODES, TRANSISTORS, AND INTEGRATED CIRCUITS

The components we have discussed so far have the same effect on voltage and current no matter how large or small or in what direction. There are other types of components whose response depends on the direction and value of voltage and current. These are usually constructed from a *semiconductor* material, such as silicon, that has been engineered to conduct in specific ways.

Semiconductors don't conduct electricity quite as well as a metallic conductor, but they do have the useful property that adding small amounts of certain impurities, called *doping*, changes their ability to conduct current. Furthermore, when adjacent areas of one piece of semiconductor are doped with different impurities, it conducts better in one direction than the other!

A component that only allows current flow in one direction is called a *diode*. Heavy-duty diodes that can withstand large voltages and currents are called *rectifiers*. If an ac current is applied to a diode, the result is a uni-directional dc current, even though the

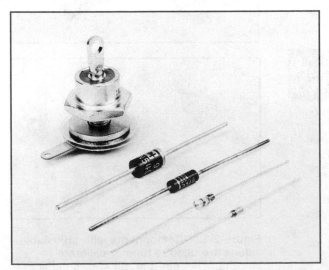
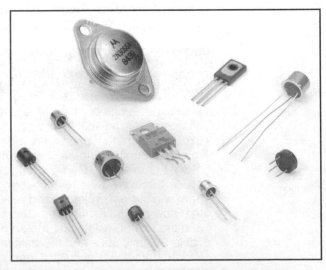

Figure 2-13—Diodes (left) and transistors (right) come in a variety of body styles. The smaller types are used in low-power circuits to control very small signals. The larger types are used for controlling power and in transmitting circuits.

current stops when it tries to flow in the "wrong" direction. **Figure 2-13** shows several types of semiconductor components.

Transistors are semiconductor material doped in patterns designed so that small voltages or currents applied at the proper point in the pattern controls larger voltages and currents. With the appropriate external circuit and a source of power, transistors can amplify or switch voltages and currents. Components that require a power source to perform their function are called *active components*. Components like resistors that don't need any power to work are called *passive components*.

An *integrated circuit* (IC or *chip*) is made of many passive and active components connected together as a useful circuit and packaged as a single component. ICs are typically constructed from a single piece of semiconductor doped in a complex pattern that forms the circuit. ICs range from very simple circuits consisting of a few diodes all the way to complex microprocessors or signal-processing chips with many thousands of active and passive components—all made from a single piece of semiconductor material.

PROTECTIVE COMPONENTS

Protective components such as those in **Figure 2-14** are used prevent equipment damage or safety hazards such as fire or electrical shock due to equipment malfunction. They are designed to have little or no effect on circuit behavior until the dangerous condition occurs. It is important to understand the different types of protective components and use them correctly.

Figure 2-14—Fuses and circuit breakers protect equipment by interrupting the current in case of an overload. Fuses "blow" by melting a metal wire or strip, seen in glass tube of the cartridge models. Circuit breakers can be reset once the problem creating the overload

Fuses interrupt excessive current flow by melting a short length of metal. When the metal melts or "blows", the current path is broken. Fuses are rated by the maximum current they can carry without blowing. Fuses can not be reused. "Slow-blow" fuses can withstand temporary overloads, but will blow if the overload is sustained.

Circuit breakers act like fuses by *tripping* when current overloads occur. Tripping opens or *breaks* the circuit. Unlike fuses, circuit breakers can be *reset* once the current overload is removed, closing the circuit and allowing current to flow again. Used in ac power wiring, a *ground-fault interrupter* (GFI) circuit breaker shown in **Figure 2-15** trips if an imbalance is sensed in the currents carried by the hot and neutral conductors. Current imbalances indicate the presence of an electrical shock hazard.

When replacing a fuse or circuit-breaker, be sure to use the same model and current rating to avoid creating a safety hazard. Using a larger value, even temporarily, could allow the fault to permanently damage the equipment or start a fire. Resist the temptation to use a higher-current device, even "just for a minute."

Surge protectors limit temporary voltage *transients* above normal voltages by turning into resistors when

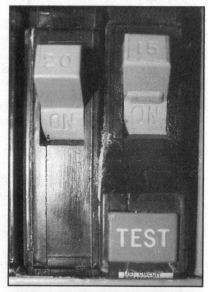

Figure 2-15—The Ground-Fault Interrupter (GFI) circuit breaker interrupts current flow when it senses imbalances between the hot and neutral circuits in ac wiring. The imbalance of currents indicates that a shock or other safety hazard exists in wiring supplied by the breaker.

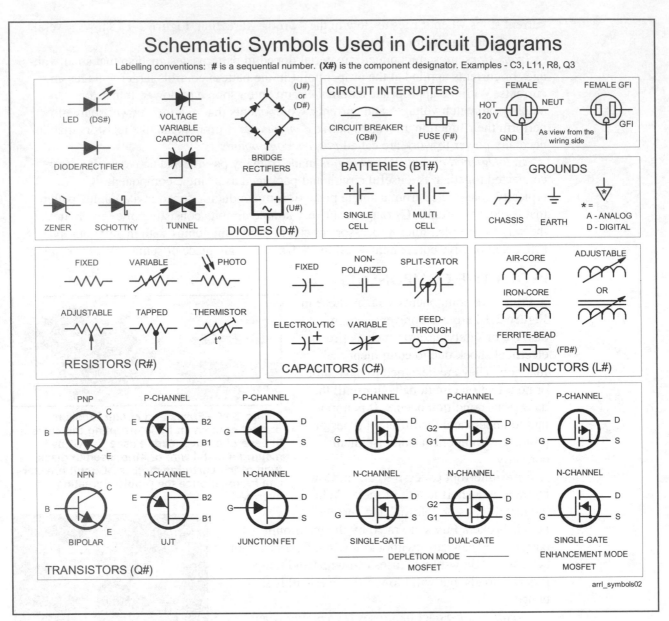

Schematic Symbols Used in Circuit Diagrams

Labelling conventions: # is a sequential number. (X#) is the component designator. Examples - C3, L11, R8, Q3

Figure 2-16—Symbols are used when drawing a circuit because there are so many types of components. Radio and electrical designers use them as a convenient way of describing a circuit.

Well-Grounded Symbols

Figure 2-16 includes three different symbols for "Grounds" which can be confusing. Remember that in the previous chapter "ground" meant a reference voltage at the earth's surface. The middle ground symbol in Figure 2-16 represents a connection directly to the earth. Not all circuits require such a reference voltage. For some equipment, the reference voltage is obtained by connecting to the metal enclosure or *chassis*, the left-hand ground symbol. The chassis may or may not have its own connection to an earth ground.

Within a piece of equipment the triangular ground symbol on the right generally indicates a connection for current to flow back to the power supply. This is usually referred to as circuit common or just common. If an A or D is added to the triangle, the circuit has digital (D) computing components (such as microprocessors) as well as circuits that handle linear signals (A). By keeping the return connection to the power supply separate for each type of circuitry, the linear signals are kept free of noise from the digital circuits.

the voltage gets too high. They then dissipate as heat the energy that would otherwise be passed on to the equipment. Surge protectors are connected to a home's ac power circuits, often in power outlet strips, and to telephone lines. *Lightning arrestors* have a similar function, but are designed to handle the much higher voltages and currents of a lightning strike.

SCHEMATICS & COMPONENT SYMBOLS

If a circuit contains more than two or three components, trying to describe it clearly in words becomes very difficult. To describe complicated circuits, engineers have developed the *schematic diagram* or just plain *schematic*. Schematics use *circuit symbols* to create a visual description of a circuit. **Figure 2-16** shows the symbols for a number of common components. Don't worry that you aren't familiar with all of them! Look for and identify the circuit symbols for a resistor, capacitor, inductor, diodes, and transistors.

Figure 2-17 shows the schematic for a simple transistor circuit. Each component is assigned a unique *designator* within the circuit or a text label; such as R1, C2, or "NPN Transistor." Resistors are designated with an R, capacitors with a C, inductors with an L, diodes with a D, and so forth. The lines between components, such as between R2 and B1,

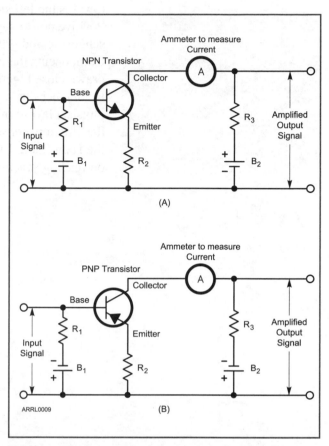

Figure 2-17—A schematic describes complex circuits using symbols representing each type of component. Lines and dots show electrical connections between the components, but may not correspond to actual wires.

Table 2-2
Electrical Units and Their Namesakes

Unit	Measures	Named for
Ampere	Current	Andree Marie Ampere 1775 -1836
Coulomb	Charge	Charles Augustin Coulomb 1736-1806
Farad	Capacitance	Michael Faraday 1791-1867
Henry	Inductance	Joseph Henry 1797-1878
Hertz	Frequency	Heinrich Hertz 1857-1894
Ohm	Resistance	George Simon Ohm 1787-1854
Watt	Power	James Watt 1736-1819
Volt	Voltage	Alessandro Giuseppe Antonio Anastasio Volta 1745 -1827

represent electrical connections. Each line does not necessarily correspond to a physical wire—it just indicates that an electrical connection exists between whatever is at each end of the line. Shared connections are shown as dots where two lines intersect. If two lines cross without a dot, there is no connection. No dots are used at the connection to a component unless more than one line is present at the connection, such as at the top end of R1.

On a well-constructed schematic, inputs to the circuit are located towards the left side of the schematic and outputs are towards the right. Positive power supply voltages are located towards the top of the schematic and ground or negative supply voltages are at the bottom. Components that work together to perform a single function are usually drawn close together. Labels are often added to indicate circuit function. Remember that a schematic may have little resemblance to the actual physical layout of the circuit. It is just a convenient way to describe how the circuit is constructed electrically. The "First Steps in Radio" link on the Ham Radio License Manual Web page will take you to a good article on reading schematics.

Before you go on, study test questions T4E07-11; T0A04, 05. Review this section if you have difficulty.

2.4 Signals and Waves

Congratulations! You have learned about the basic types of radio equipment and the fundamentals of electronics. Now we can start learning about radio signals—real radio! Communicating by radio depends on the processes of modifying a signal's characteristics, adding and subtracting them, making them larger and smaller, and using those characteristics to represent information such as speech or text or data. We'll start at the beginning!

FREQUENCY & PHASE

One of the most fundamental characteristics of any ac signal, radio or otherwise, is its *frequency*. A complete sequence of ac current flowing, stopping, reversing, and stopping again is called a *cycle*. The same is true for an ac voltage building up to a positive voltage from zero, then reversing to negative polarity, and returning to zero. The most common shape of ac current or voltage is a *sine wave* of only one frequency as shown in **Figure 2-18**.

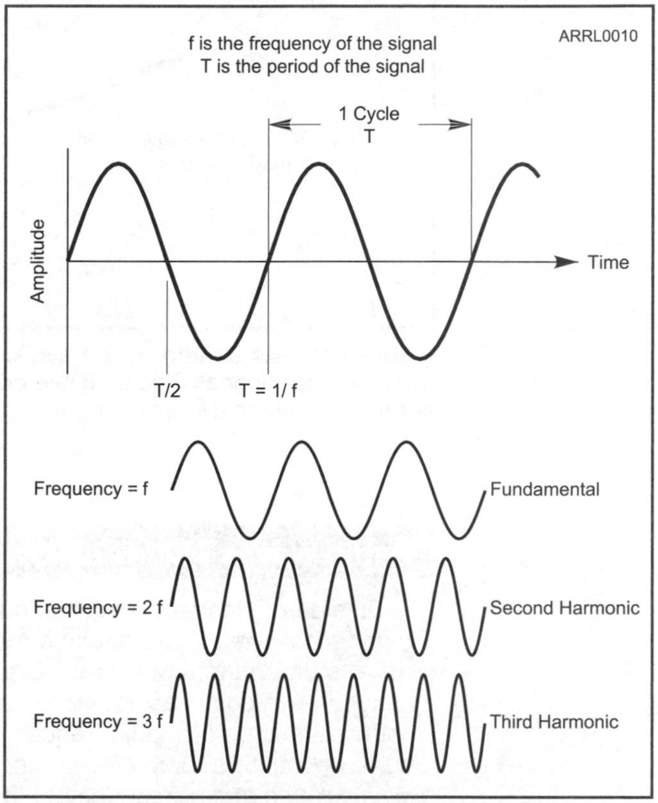

The number of cycles per second is the signal's frequency, represented by a lower-case letter f. The unit of measurement for frequency is *Hertz*, abbreviated Hz. One cycle per second is one hertz or 1 Hz. The duration or *period* of the cycle (represented by capital T) is fixed. The reciprocal of the period, 1/T, is the signal's frequency, f.

A *harmonic* is an RF signal at a frequency that is some integer multiple (2, 3, 4, etc.) of a lowest or fundamental frequency. The harmonic at twice the fundamental's frequency is called the *second harmonic*, at three times the fundamental frequency the *third harmonic*, and so forth. There is no "first harmonic."

Figure 2-18—The frequency of a signal and its period are reciprocals. Higher frequency means shorter period and vice versa. Harmonics are signals with frequencies that are integral multiples of a fundamental frequency.

WAVELENGTH

The *wavelength* of a radio wave is the distance that it travels during one complete cycle. Wavelength is represented by the Greek letter lambda, λ. **Figure 2-19** shows that wavelength is equal to the speed of the wave divided by its frequency. All radio waves travel at the speed of light (represented by a lower-case c) in whatever medium they are traveling, such as air or the vacuum of space. The speed of light in space and air is 300 million meters per second. In water, along wires, and inside cables, c is lower by up to one-third. Even though its frequency is unchanged, because the wave is moving slower, its wavelength becomes shorter.

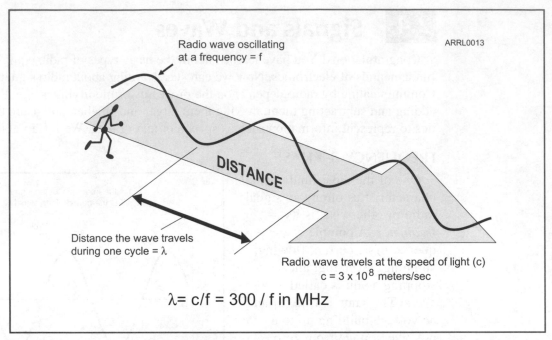

Figure 2-19—As a radio wave travels, its electric and magnetic fields oscillate at the frequency of the signal. The distance covered by the wave during the time it takes for one complete cycle is its wavelength.

Radio Waves - On Their Own

Outside of electronic circuits and wires, radio signals take the form of *electromagnetic waves*. That's surely a mouthful, but all it means is that the energy in a *radio wave* is divided into two forms. One form is an *electric field* that creates a voltage between two points as the wave passes by. The other form is a *magnetic field*. Both combine to create a radio frequency ac current in any conducting material that the wave encounters. Both electric and magnetic fields are constantly changing directions or *oscillating* at the frequency of the signal that was used to generate the wave.

A radio wave can be referred to by wavelength or frequency because the two are related by the speed of light. Because radio waves travel at a constant speed (in one media), one wavelength, $\lambda = c / f$. This can also be stated as $f = c / \lambda$. For waves in air or space, the formula for wavelength in meters is:

$\lambda = 300,000,000$ / frequency in Hertz

$\lambda = 300 / f$ in MHz

For example, the wavelength of a 1 MHz radio wave (typical of an AM broadcast station) is:

$\lambda = 300,000,000 / 1,000,000 = 300$ meters

$\lambda = 300 / 1 = 300$ meters

Clearly, the second form is more convenient when working with RF signals. To convert from meters to feet, multiply the wavelength in meters by 3.28. To convert from meters to inches, multiply by 39.37.

RESONANCE

In a circuit with both capacitive and inductive reactance, at some frequency the two types of reactance will be equal. At that frequency, the effects of the capacitor and inductor cancel. The ac current and voltage are brought exactly back in step with each other—a condition called *resonance*. The frequency at which resonance occurs is the *resonant frequency*.

When a circuit is *resonant*, opposition to the flow of current, ac or dc, is as if only resistance was present—no reactance. At resonance, impedance is said to be *purely resistive*. Depending on how the circuit components are connected, the cancellation of capacitive and inductive reactance at resonance can result in very high or very low impedance to ac signals.

If a resonant circuit is used to optimize a circuit's performance, that is a *tuned circuit*. If variable capacitors or inductors are used, the resonant frequency of the circuit can be varied. By placing tuned circuits at the right point in a circuit, they can be used to block or pass ac signals.

Signals that have a frequency greater than 20,000 Hz (or 20 kHz) are *radio frequency* or RF signals. The range of frequencies of RF signals is called the *radio spectrum*. It starts at 20 kHz and continues through several hundred GHz, a thousand million times higher in frequency! Signals below 20 kHz are *audio frequency* or AF signals.

For convenience, the radio spectrum of **Figure 2-20** is divided into ranges of frequencies that have similar characteristics as shown in **Table 2-3**. Frequencies above 1 GHz are

Table 2-3
RF Spectrum Ranges

Range Name	Abbreviation	Frequency Range
Very Low Frequency	VLF	3 kHz - 30 kHz
Low Frequency	LF	30 kHz - 300 kHz
Medium Frequency	MF	300 kHz - 3 MHz
High Frequency	HF	3 MHz - 30 MHz
Very High Frequency	VHF	30 MHz - 300 MHz
Ultra High Frequency	UHF	300 MHz - 3 GHz
Super High Frequency	SHF	3 GHz - 30 GHz
Extremely High Frequency	EHF	30 GHz - 300 GHz

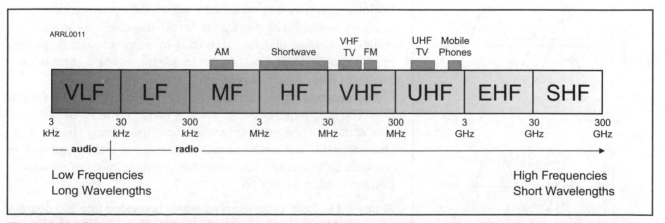

Figure 2-20—The radio spectrum extends over a very wide range of frequencies. The drawing shows the frequency ranges used by broadcast and mobile phones. Amateurs can use small frequency bands in the HF and higher frequency regions of the spectrum.

Filters - Radio Gatekeepers

Filters perform very important functions in radio, as you will see in Chapter 3. Just like a filter you might use to remove dust or impurities from a stream of air or water, radio filters reduce the strength of unwanted signals. This reduction is called *attenuation*. There are many types of filters, but the most common in ham radio are the *low-pass*, *high-pass*, *band-pass*, and *notch*. These terms refer to how filters treat signals of different frequencies. For example, in a low-pass filter, low frequencies pass through at full strength, but high frequencies are attenuated. Band-pass filters pass a band of frequencies. A notch filter is like a band-pass filter in reverse—a band of frequencies is attenuated while all others pass. Band-pass and notch filters are *wide* or *narrow* depending on whether the frequency range of interest is large or small. You'll meet all of these filters in person later!

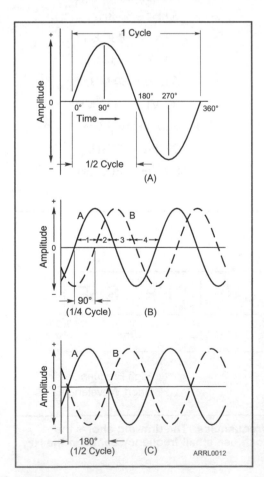

generally considered to be *microwaves*. Microwave ovens operate at 2.4 GHz, for example. Hams primarily use frequencies in the MF through UHF and microwave ranges. A specific range of frequencies used for a common purpose or with common characteristics is called a *band*. Frequency bands used by amateurs are called *amateur bands* or *ham bands*.

Let's return to the sine wave signal of Figure 2-18. Every cycle of the signal has the same basic shape; rising and falling and returning to where it started. Position within a cycle is called *phase*. Phase is used to compare how sine wave signals are aligned in time.

Phase is measured in *degrees* or *radians*. There are 360 degrees or 2π radians in one cycle of a sine wave. (The Greek letter pi, π, is pronounced "pie" and represents the value 3.1416). If two sine waves have a phase difference of 180 degrees or π radians so that one wave is increasing while the other is decreasing, they are *out of phase*. Waves that have no phase difference so that they are increasing and decreasing at the same time are *in phase*. This is illustrated in **Figure 2-21**.

MODULATION

A simple radio signal just sitting there on the radio spectrum not doing anything isn't very useful and doesn't do much communicating. To communicate, information must be added or contained in the radio signal. How does that happen?

The simplest radio signal at one frequency whose strength never changes is called a *continuous wave*, abbreviated CW. Adding information to a signal by modifying it in some way is called *modulation*. The combination of modulation method and type of information is the signals *mode*. Recovering the information from a modulated signal is called *demodulation*. A signal that doesn't carry any information is called *unmodulated*.

The simplest mode is a continuous wave turned on and off in a coded pattern, such as the Morse code. In fact, Morse code signals are often called CW for that reason. The official name for Morse code modulation is *radiotelegraphy*. If speech is the information used to modulate a signal, the result is a *voice mode* signal. If data is the information used to modulate a signal, the result is a *data mode* or *digital mode* signal. *Analog* modes are those which carry information that can be understood directly by a human such as speech or Morse code. Digital or data modes are those which carry information as data characters between two computers.

Any characteristic of a signal can be varied to contain information. The only requirement is that the variations must be observable at the receiving end to recover the information. The three characteristics that can be modulated are the signal's *amplitude* or strength, its frequency, and its phase. All three types of modulation are used in ham radio.

Figure 2-21—Each cycle of an ac signal is divided into 360 degrees or 2π radians of phase (a). Phase is used as a measure of time within the signal. Parts (b) and (c) show two special cases; in (b) the two signals are 90 degrees out of phase and in (c) they are 180 degrees out of phase.

AMPLITUDE MODULATION

Varying the power or amplitude of a signal to add speech or data information is called *amplitude modulation* or AM. (Morse code is the simplest form of AM.) If you have watched a meter jump in response to your voice or music, an AM signal's amplitude has to change in the same way to carry the information in your voice. An AM transmitter adds your voice to the unmodulated signal by varying its amplitude just as the jumping meter suggests. The information is contained in the outline or *envelope* of the resulting signal. **Figure 2-22** shows the result of using a tone to create an AM signal.

All a receiver has to do to recover your voice from an AM signal is to follow the signal's amplitude variations and your voice reappears! The process of recovering speech or music by following the envelope of an AM signal is called *detection* and can be performed by very simple circuits. In fact, the first AM receivers were nothing more than a crystal of galena (lead ore), a thin steel sliver (called a "cat's whisker"), and a sensitive pair of headphones! AM is widely used because it is simple to transmit and receive.

An actual AM signal is really a composite signal—three separate signals working together. These components are a *carrier* and two *sidebands*. The total power of an AM signal is divided between the three components. If you could "look" closely at an AM signal containing the information of a single, steady tone, such as the emergency broadcast tests heard on AM radio, you would see the three separate signals—carrier and sidebands—very close to each other. You would notice that the AM signal's carrier is a continuous wave whose amplitude does not change. It contains no information and is called a carrier because it "carries" the information that the receiver then detects.

An AM signal carrying a tone has two sidebands that are present as steady, unchanging signals as long as the tone is transmitted. The *upper sideband* or USB is higher in frequency than the carrier by the frequency of the tone. The *lower sideband* or LSB is lower in frequency than the carrier. The information to recover the tone is contained in the amplitude of the sidebands and its difference in frequency from the carrier.

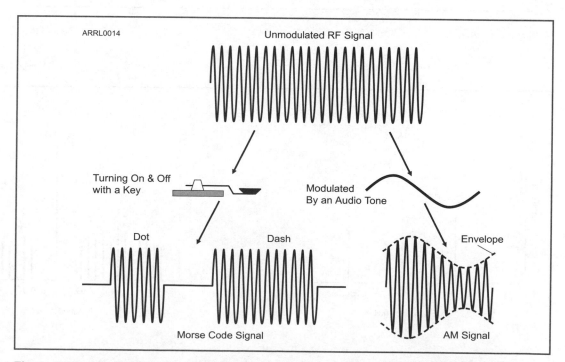

Figure 2-22—Information can be added to an RF signal by modulating the signal's amplitude. Turning the signal on and off in a pattern such as the Morse code is a very simple form of amplitude modulation. A tone or speech can also be used to modulate the signal, resulting in a signal whose shape contains the information from the tone or speech.

Can AM Carry a Tune?

Figure 2-23 shows a new way of looking at signals. Instead of showing how the signal amplitude varies with time from left to right, the graph shows how signals are spread out in frequency. This is how your AM broadcast receiver "sees" the radio spectrum. With lower frequencies on the left, as you tuned the receiver higher in frequency, it would first encounter the signal at left, tune "past" it, then encounter the next signal, and so forth. This is called a *frequency domain* graph because the horizontal scale is frequency, not time. The vertical scale still represents the signal amplitude. Graphs of signal amplitude variations against time like those of the sine wave show the *time domain*.

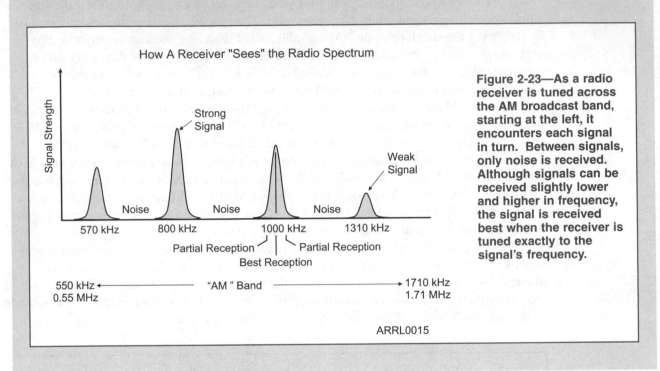

Figure 2-23—As a radio receiver is tuned across the AM broadcast band, starting at the left, it encounters each signal in turn. Between signals, only noise is received. Although signals can be received slightly lower and higher in frequency, the signal is received best when the receiver is tuned exactly to the signal's frequency.

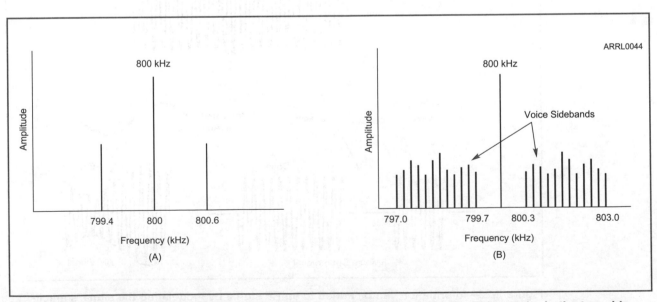

Figure 2-24—At (a), the 800 kHz carrier is modulated by a single tone of 600 Hz. This results in the two sidebands 600 Hz away from the carrier. At (b) the many frequencies present in a voice signal from 300 to 3000 Hz are represented as many smaller sidebands, each corresponding to a part of the voice signal.

Composite and Complex Signals

Composite signals, both radio and audio, are groups of individual signals that combine to create a complex signal, such as one that carries a tone shown in Figure 2-24. Composite signals have *components* that may cover a range of frequencies. The difference in frequency between the lowest and highest component of a composite signal is the signal's *bandwidth*. CW signals occupy the least bandwidth, followed by SSB, AM and FM.

Figure 2-24(A) shows an AM signal with a carrier of 800 kHz modulated by a single, steady tone of 600 Hz. As your receiver tuned to the AM signal from the left, it would first encounter the lower sideband as a steady signal at 799.4 kHz, then the carrier at 800 kHz, 600 Hz higher in frequency. The carrier would be twice as strong as the sideband. Continuing up in frequency, the upper sideband would be encountered 600 Hz above the carrier at 800.6 kHz. This set of three component signals combine to make one AM signal. Both sidebands each contain the information needed to reproduce the tone used to modulate the signal.

Figure 2-24 also shows a picture of an AM signal carrying the information in a complex voice signal. Because voice or music contains many frequencies and the amplitude is constantly changing, an AM signal carrying that information looks much more complex, although the carrier remains steady.

From the standpoint of power, an AM signal is inefficient. First, the carrier doesn't carry any information, even though it consumes most of the signal power! Then, each sideband contains an exact copy of the modulating signal. It seems like only a fraction of the signal is really needed and that's just what *single-sideband* or SSB signals are. **Figure 2-25** shows a single-sideband mode signal; an AM signal with the carrier and one sideband removed electronically. All of the SSB signal's power can then be devoted to the remaining sideband. The upper sideband (USB) is used on VHF and UHF.

Even though generating and receiving SSB signals requires more complex equipment, the performance improvement is worth the trouble. SSB transmissions have a superior range compared to AM because all of the power is concentrated in a single, information-carrying sideband. SSB's smaller bandwidth of 2-3 kHz also makes better use of the radio spectrum because more SSB signals can fit in a fixed range of frequencies without overlapping or interfering with each other.

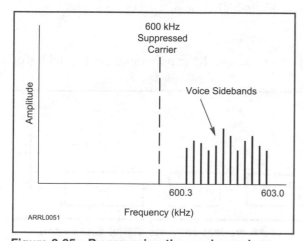

Figure 2-25—By removing the carrier and one sideband, a single-sideband (SSB) signal is created. All of the signal's power is concentrated in one sideband. SSB signals are used for long-distance and weak signal voice contacts.

FREQUENCY & PHASE MODULATION

The remaining two signal characteristics used to carry information are frequency and phase. Modes that vary the frequency of a signal to add speech or data information are called *frequency modulation* or FM. The frequency is varied in proportion to the strength of the modulating signal as shown in **Figure 2-26**. For example, speaking louder into the microphone of an FM transmitter increases the amount of frequency variation. The amount that an FM signal's frequency varies when modulated is called *deviation*.

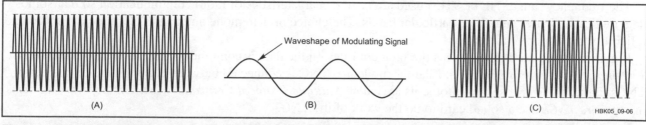

Figure 2-26—At (a), each cycle of the unmodulated carrier has the same frequency. When the carrier is frequency-modulated with the signal at (b), its frequency increases and decreases corresponding to the increases and decreases in amplitude of the modulating signal.

Phase modulation or PM is closely related to FM. Phase modulation is created by varying the amount a signal is delayed instead of changing its frequency. These two techniques result in transmitted signals that are approximately the same. Receivers can demodulate FM and PM with the same demodulator circuit. Most amateurs refer to either FM or PM signals simply as FM.

FM signals have one carrier and many sidebands that all add together in a 5 to 15 kHz bandwidth. The amplitude of an FM signal is not changed by the modulating information, so an FM signal always has the same power, whether modulated or not. This is called a *constant power* signal.

If compared to SSB, FM signals occupy more bandwidth, why is FM used so widely for VHF and UHF voice repeaters? Recall that the information in an FM signal is entirely contained in the signal's frequency variations. Atmospheric and electrical noises are received as amplitude variations, but are meaningless to an FM signal. The *limiter* circuit in an FM receiver strips away all of the amplitude variations from FM signals, including noise, so that it is not heard in the receiver's output. That's why programs on AM stations experience static crashes at the same time an FM station's program is quiet. For short-range and regional communications where high-power transmitters are not required, FM delivers higher quality reception.

DATA MODES

Amateurs have developed or adapted techniques for exchanging digital data by transforming the 1's and 0's of digital data into tones that are in the same frequency range as the human voice. Radios designed for voice transmission then transmit and receive the tones as either AM or FM signals. A *modem*, short for modulator-demodulator, changes data signals to and from audio signals. Modems may be standalone devices built into a radio or computer software and a sound card. **Figure 2-27** shows how a computer, modem, and radio are combined to communicate using digital data.

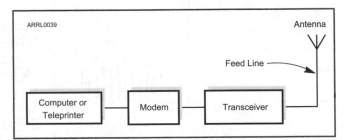

Figure 2-27—A modem converts digital data to and from audio tones that can be communicated by radio. When a microprocessor that implements a digital data protocol is added to a modem, the result is a Terminal Node Controller or TNC.

Just changing the data to tones is not enough for reliable communications. A radio signal experiences many disruptions between the transmitter and receiver; fading, interference, noise. These would cause errors in the recovered data if no precautions were taken. The data must be arranged in such a way that helps the receiver detect and correct those errors. There must also be a process by which the transmitter and receiver coordinate the exchange of data. The rules that describe the arranging and exchange of data are called a *protocol*.

When a protocol is combined with a modulation method such as SSB or FM, a *data mode* is created. Examples of data modes include:
- Radioteletype (RTTY) that uses the 5-bit Baudot protocol and SSB modulation
- VHF Packet radio that uses the AX.25 protocol and FM modulation
- Winlink 2000 that uses the B2F protocol and SSB modulation
- Keyboard-to-keyboard modes that combine protocols such as PSK31,MFSK, AMTOR, and PACTOR with SSB modulation

The frequency range (HF or VHF) used affects the characteristics of radio communication so that some data modes are better suited for particular bands. The choice of data mode and band determine the speed and reliability of data communications.

If a microprocessor that performs the protocol rules on the data is combined with a modem, the result is a *Terminal Node Controller* (TNC). Like the modem, a TNC is connected between a computer and the radio. TNCs that can perform several protocols are called *Multiple Protocol Controllers* (MPC). *Software modems* or *software TNCs* use a sound card to do the work of the TNC.

Before you go on, study test questions T4A05; T4B01-09; T6A01, 02, 04-10; T9A05. Review this section if you have difficulty.

2.5   Antennas & Feed lines

No piece of equipment has as great an effect on the performance of a radio station, whether hand-held or home-based, as the antenna. Experimenting with antennas has been a favorite of hams from the very beginning, contributing greatly to the development of antennas for all radio services. For these reasons, knowledge of antennas is very important for amateurs.

ANTENNA FUNDAMENTALS

(a)

(b)

Figure 2-28—The ground-plane antenna shown at (a) radiates a signal from the vertical wire attached to the base. The vehicle's metal surface acts an electrical mirror, creating the effect of another wire opposite to the one above the surface. At (b) the Yagi-style beam antenna uses parasitic elements to direct the signal in one direction and reject signals in the opposite

An antenna is not necessarily something that you must purchase or that "looks like" a commercial product. Any electrical conductor can act as an antenna for radio signals—a wire, a pipe, a car body, even the proverbial bedsprings. However, for an antenna to radiate and receive radio signals efficiently, its dimensions must be an appreciable fraction of the signal's wavelength. **Figure 2-28** shows two examples of antennas commonly used by Technician class hams.

A *feed line* is used to deliver the radio signals to or from the antenna. The connection of antenna and feed line is called the *feedpoint* of the antenna. The ratio of radio frequency voltage to current (just like Ohm's Law) at an antenna's feedpoint is the antenna's *feedpoint impedance*. An antenna is resonant when its feedpoint impedance is all resistance with no reactance.

An antenna's feedpoint impedance depends on how its physical dimensions compare to the wavelength of radio signals at that frequency. Feedpoint impedance changes with frequency because wavelength changes but the physical dimensions don't. An antenna's feedpoint impedance is also affected by nearby conductors and its height above ground.

The conducting portions of an antenna by which the radio signals are transmitted or received are called *elements*. An antenna with more than one element is called an *array*. The element connected to the feed line is called the *driven element*. If all of the elements are connected to a feed line, that's a *driven array*. Elements that are not directly connected to a feed line, but that influence the antenna performance are called *parasitic elements*.

Polarization refers to the direction in which the electric field of a radio wave is oriented. A *horizontally polarized* antenna has elements aligned parallel to the surface of the earth and radiates a radio wave whose electric field is oriented horizontally. A *vertically polarized* antenna has elements and a radiated electric field perpendicular to the surface of the earth.

Gain

The concentration of radio signals in a specific direction is called *gain*. An antenna can create gain by radiating radio waves that add together in the preferred direction and cancel in others. Gain can also be created by reflecting radio waves so that they are focused in one direction. Gain increases signal in the preferred direction for both receiving and transmitting. Gain only focuses power, it does not create power. Gain aids communication in the preferred direction by increasing transmitted and received signal strengths.

An *isotropic* antenna has no gain because it radiates equally in every possible direction. Isotropic antennas do not exist in practice and are only used as an imaginary reference. An *omnidirectional* antenna radiates a signal equally in every horizontal direction. An antenna with gain ina single direction is called a *beam* or *directional* antenna. Omnidirectional antennas are useful for communicating over a wide region while beam antennas are used when communication is desired in one direction.

An antenna's gain is specified in decibels (dB) with respect to an identified reference antenna. For example, the abbreviation dBi means gain in decibels with respect to an isotropic antenna. dBd means gain with respect to a dipole antenna. (Dipole antennas are covered in Chapter 3.) Gain, like voltage, is a relative measurement between an antenna and some reference, such as an isotropic or dipole antenna.

RADIATION PATTERNS

The easiest way to describe how an antenna distributes its signals is a graph showing the antenna's gain in any direction around the antenna. That graph is called a *radiation pattern*. An antenna transmits and receives with the same pattern.

The most common type of radiation pattern is an *azimuthal* pattern that shows the antenna's gain in horizontal directions around the antenna. An azimuthal pattern can be imagined as looking down on the antenna from above as in **Figure 2-29**. An *elevation* pattern shows the strength of the radiated energy in vertical directions as if the antenna is viewed from the side, as shown in **Figure 2-30**, or from the front. An antenna's radiation pattern may change as frequency changes for the same reasons that feedpoint impedance changes with frequency— changes in frequency change the wavelength while the physical dimensions remain fixed.

The region of the radiation pattern in which the antenna's gain is greatest is called the *main lobe*. Regions of less gain are called *side lobes* and *nulls* where gain is a minimum. The ratio of gain in the preferred or *forward* direction to that in the opposite direction is called the *front-to-back ratio*. The ratio of gain in the forward direction to that at right angles called the *front-to-side ratio*. Antennas with high front-to-back and front-to-side ratios are useful in rejecting interference and noise from unwanted directions.

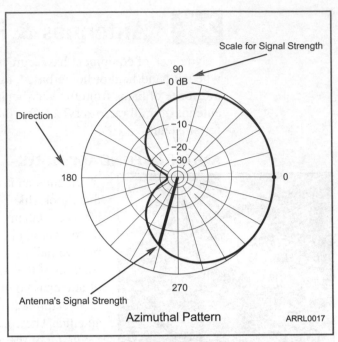

Figure 2-29—As if looking down on the antenna from above, the azimuth radiation pattern shows how well the antenna transmits or receives in all horizontal directions. The distance from the center of the graph to the solid line is a measure of the antenna's ability to receive or transmit in that direction.

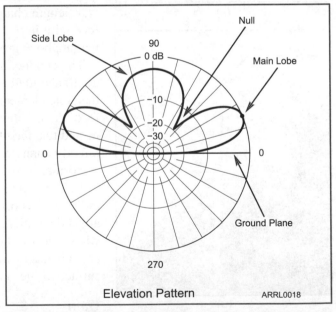

Figure 2-30—The elevation pattern looks at the antenna from the side to see how well it receives and transmits at different angles above the plane of the earth.

FEED LINES & SWR

Feed lines are made from two conductors separated by an insulating material such as plastic. The radio signal is carried on the conductors and in the space between them. Feed lines used at radio frequencies use special materials and construction methods to avoid *feed line loss*; leaking radio signals or dissipating them as heat.

The most popular feed line used by amateurs is *coaxial cable* or *coax*. **Figure 2-31** shows how coaxial cable is constructed. A wire *center conductor* is surrounded by insulation (the *center insulator*). The insulation is covered with a tubular *shield* of braided wire or foil. Finally, the cable is covered with a plastic sheath called the *jacket*. Coaxial cable carries the radio signal between the center conductor and the inside surface of the shield. That means it can be placed next to other conducting surfaces such as conduit or antenna support masts without affecting the signal inside.

A feed line of two parallel wires separated by insulating material is called *open-wire*, *ladder line*, or *twinlead*. This type of feed line, also shown in **Figure 2-31**, has less insulating material for the radio energy to move through, so it has less loss than coaxial cable. Since the radio energy is not shielded by an outer tube, the signals in twinlead can be affected by nearby conductors. Twinlead feed lines can not be used in enclosed conduits and must be kept clear of nearby conducting surfaces.

Feed lines have a *characteristic impedance*, abbreviated as Z_0, which is a measurement of how energy is carried by the feed line. This is not the same as the resistance of the conductors if measured from end to end of the feed line. To understand characteristic impedance, try this experiment: Obtain thin and fat drinking straws. Blow a single puff of air through each, feeling the resistance to air flow through each. You will feel more back-pressure in the thin straw because to the flow of air its characteristic imped-

Figure 2-31—This drawing illustrates some common types of open-wire and coaxial cables used by amateurs. Open-wire line has two parallel conductors separated by insulation. "Coax" has a center conductor surrounded by insulation. The second conductor, called the shield, covers the insulation and is, in turn, covered by the plastic outer jacket.

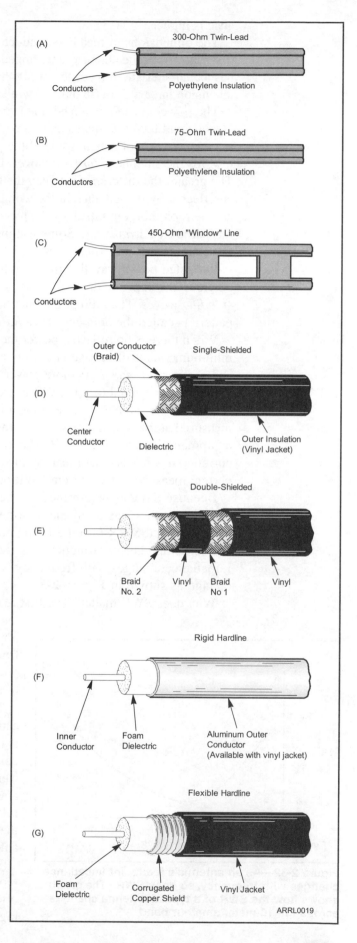

(A) 300-Ohm Twin-Lead
Conductors
Polyethylene Insulation

(B) 75-Ohm Twin-Lead
Conductors
Polyethylene Insulation

(C) 450-Ohm "Window" Line
Conductors

(D) Single-Shielded
Outer Conductor (Braid)
Center Conductor
Dielectric
Outer Insulation (Vinyl Jacket)

(E) Double-Shielded
Braid No. 2
Vinyl
Braid No 1
Vinyl

(F) Rigid Hardline
Inner Conductor
Foam Dielectric
Aluminum Outer Conductor (Available with vinyl jacket)

(G) Flexible Hardline
Foam Dielectric
Corrugated Copper Shield
Vinyl Jacket

ARRL0019

ance is higher.

The dimensions of feed line conductors, the spacing between them, and the insulating material determine characteristic impedance. Most coaxial cable used in amateur radio has a characteristic impedance of 50 ohms. Coaxial cables used for video and cable television have a Z_0 of 75 ohms. Twinlead feed lines have a Z_0 of 300 to 450 ohms.

The power carried by a feed line is transferred completely to an antenna when the antenna and feed line impedances are identical or *matched*. If the feed line and antenna impedances do not match, some of the power is *reflected* by the antenna. Power traveling towards the antenna is *forward power*. Power reflected by the antenna is *reflected power*. The greater the difference between the feed line and antenna impedances, the more power is reflected by the antenna. In the worst case where the feed line ends or is *terminated* in an open- or short circuit, all of the forward power is reflected. Power in a feedline is measured by a *wattmeter*. Some wattmeters can read both forward and reflected power—these are *directional wattmeters*.

Reflected power traveling in a feed line creates an interference pattern with forward power in the opposite direction. The pattern is stationary along the feed line and is called *standing waves*. The ratio of the maximum value to minimum value of the interference pattern is called the *Standing Wave Ratio* or SWR.

When there is no reflected power there is no interference pattern, the maximum and minimum values of voltage or current are the same and the SWR is 1:1. This condition is called a *perfect match*. As more power is reflected, SWR increases. SWR is always greater than or equal to 1:1. SWR is greater than 1:1 is called a *mismatch*.

SWR is the same anywhere along a feed line, but it is most commonly and conveniently measured at the transmitter's connection to the feed line. An *SWR Meter* is the piece of equipment that measures SWR. The interference pattern can be measured as voltages or currents. If SWR is determined by measuring voltage, it is Voltage SWR or VSWR. If by current measurements, Current SWR or ISWR. Both give the same value for SWR.

Because SWR is determined by the proportions of forward and reflected power, SWR is also a measure of the relationship between the antenna and feed line impedances. In fact, SWR is equal to the ratio of antenna-to-feed line or feed line-to-antenna impedances, whichever is greater than 1. Since an antenna's feedpoint impedance changes with frequency while that of the feed line does not, so will SWR change as shown in **Figure 2-32**.

Why does SWR matter? The first reason is that transmitters are designed to deliver the most output power when connected to a specific impedance, usually 50 ohms. As the impedance varies from 50 ohms, meaning higher SWR, less output power is available and the transmitter has to work harder to produce it.

The second reason is that the interference pattern causes voltages to increase in the feed line, including at the transmitter's output where the feed line is connected. The higher voltages caused by high SWR can damage a transmitter's output circuits. Most amateur transmitting equipment is designed to work at full power with an SWR of 2:1 or lower. SWR greater than 2:1 may cause the transmitter to reduce power automatically in order to protect its output circuit.

What causes high SWR? Antennas that are much too short or too long for the frequency being used often have extreme feedpoint impedances, causing high SWR. High SWR can also be caused by a faulty

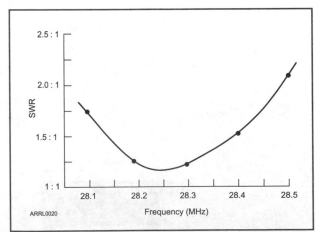

Figure 2-32—As an antenna's feedpoint impedance changes with frequency, so does SWR. The graph shows how the SWR of a typical antenna changes across the 10-meter amateur band.

feed line or feed line connectors. To correct high SWR not caused by a fault, an impedance matching device is used. This device is called a *transmatch*, an *impedance matcher*, or an *antenna tuner*. The antenna tuner is typically connected at the output of the transmitter.

An SWR meter is placed between the antenna tuner and transmitter as shown in **Figure 2-33** so that the operator can adjust the antenna tuner for minimum SWR. An antenna tuner is adjusted until the SWR measured at the transmitter output is acceptably close to 1:1. Think of the antenna tuner as an electrical gearbox that lets the engine (the transmitter) run at the speed it likes no matter how fast the tires are turning (feedpoint impedance).

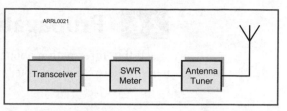

Figure 2-33—An antenna tuner is connected to the antenna feed line. The SWR meter, connected between the tuner and the transceiver, allows the operator to adjust the tuner for minimum SWR at the transceiver's output.

Before you go on, study test questions T9A03; T9B09; T9C01-06, 08, 11, 12. Review this section if you have difficulty.

2.6 Propagation

All that information about radio signals and waves was great, but how do they get from here to there? This chapter will teach you about *propagation*, which is the movement of radio waves.

BASIC PROPAGATION

Radio waves spread out from an antenna in straight lines unless reflected or diffracted along the way, just like light. Light waves are just a very, very, very high frequency form of radio waves! And like light, the strength of a radio wave decreases as it travels farther from the transmitting antenna. Eventually the wave becomes too weak to be received because it has either spread out too much or something along its path absorbed or scattered it. The distance over which a radio transmission can be received is called *range*.

If the transmitting and receiving antennas are within direct sight of each other, that's *line-of-sight propagation*. Most propagation at VHF and higher frequencies is line-of-sight. Increasing antenna height or transmitter power also increases the range of line-of-sight propagation. Radio waves at HF and lower frequencies can also travel along the surface of the earth as *ground wave propagation.*

Radio waves can be reflected by any sudden change in the media through which they are traveling, such as a building, hill, or even weather-related changes in the atmosphere. Obstructions such as buildings and hills create radio shadows, especially at VHF and UHF frequencies. **Figure 2-34** shows how radio waves can also be *diffracted* or bent as they travel past edges of buildings or hills. Radio signals arriving at a receiver after taking different paths from the transmitter can interfere with each other if they are out of phase, even canceling completely! This phenomenon is known as *multi-path*. Signals from a station moving through an area where multi-path is present have a characteristic rapid variation in strength known as *mobile flutter* or *picket-fencing*.

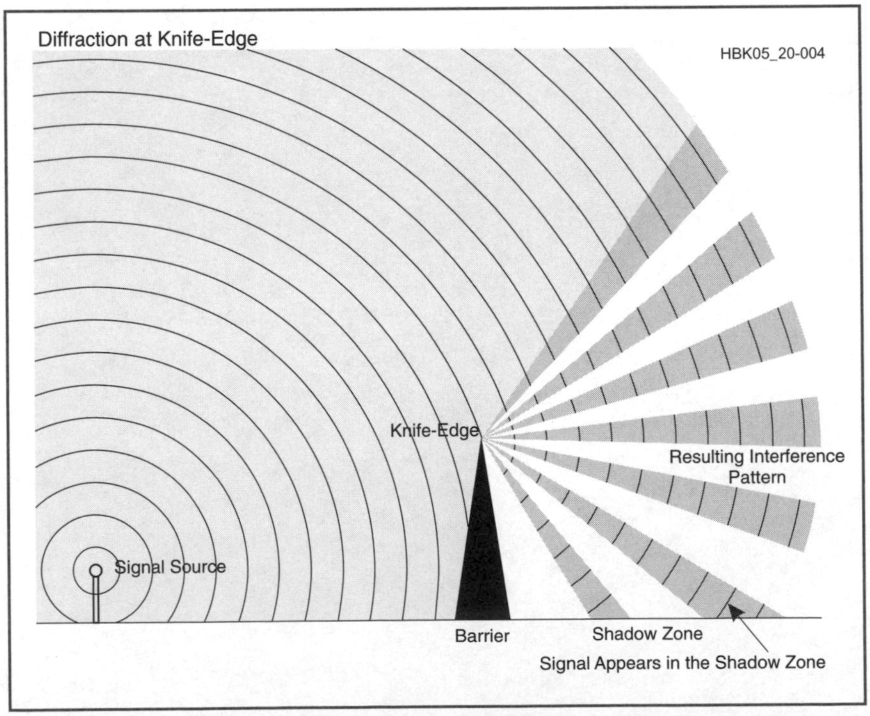

Propagation above VHF frequencies assisted by atmospheric phenomena such as weather fronts or temperature inversions is called *tropospheric propagation* or just "tropo." Radio signals are also reflected by conducting things in the atmosphere, such as planes, meteor trails, and even the aurora. All of these can reflect signals over hundreds or thousands of miles and are regularly used by amateurs to make short contacts that would otherwise be impossible by line-of-sight propagation. Yet there is one more conductive thing floating around "up there" that hams use every day to communicate around the world.

Diffraction at Knife-Edge
HBK05_20-004
Knife-Edge
Resulting Interference Pattern
Signal Source
Barrier
Shadow Zone
Signal Appears in the Shadow Zone

Figure 2-34—Radio waves are diffracted around the sharp edge of a solid object, such as a building, hill, or other obstruction. Some signals appear behind the obstruction as a result of interference between waves at the edge and those farther away. The resulting interference pattern creates shadowed areas where little signal is present.

THE IONOSPHERE

Above the lower atmosphere where the air is relatively dense and below outer space where there isn't any air at all lies the *ionosphere*. In this region from 30 to 260 miles above the earth, atoms of oxygen and nitrogen gas are exposed to the intense and energetic *ultra-violet* (UV) rays of the sun. These rays have enough energy to create positively charged *ions* from the gas atoms by knocking loose some of their negatively charged electrons. The resulting ions and electrons create a weakly conducting region high above the earth.

Because of the way the atmosphere is structured, the ionosphere forms as layers shown in **Figure 2-35**, called the D, E, F1 and F2 layers, with the D-layer being the lowest. Depending on whether it is night or day and the intensity of the solar radiation these layers can diffract (E, F1, and F2 layers) or absorb (D and E layers) radio waves. The ability of the ionosphere to diffract or bend radio waves also depends on the frequency of the radio wave. Higher frequency waves are bent less than those of lower frequencies. At VHF and higher frequencies the waves usually pass through the ionosphere with only a little bending and are lost to space.

Radio waves at HF (and sometimes VHF) can be completely bent back towards the earth by the ionosphere's F-layers as if they were reflected. This is called *sky wave propagation* or *skip*. Since the earth's surface is also conductive, it can reflect radio waves, too.

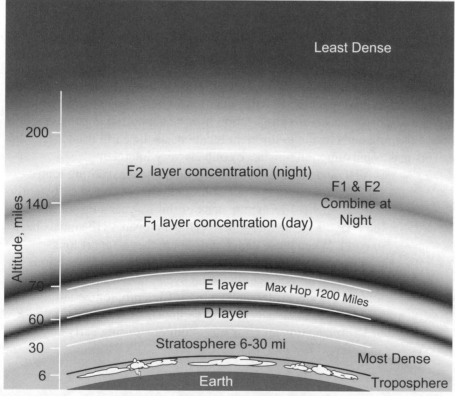

UBR1-1502

Ionosphere   30 to 260 Miles Above the Surface

Least Dense

200

F2 layer concentration (night)

F1 & F2 Combine at Night

140

F1 layer concentration (day)

Altitude, miles

70 — E layer   Max Hop 1200 Miles

60 — D layer

30 — Stratosphere 6-30 mi

6 —   Most Dense

Earth   Troposphere

Figure 2-35—The ionosphere is formed by solar ultraviolet (UV) radiation. The UV rays knock electrons loose from air molecules, creating weakly charged layers at different heights. These layers can absorb or refract radio signals, sometimes bending them back to the earth.

Figure 2-36—Signals that are too low in frequency are absorbed by the ionosphere and lost. Signals that are too high in frequency pass through the ionosphere and are also lost. Signals in the right range of frequencies are refracted back towards the earth and are received hundreds or thousands of miles away.

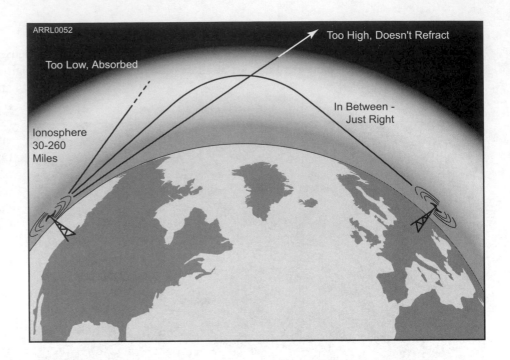

This means that a radio wave can be reflected between the ionosphere and the earth multiple times. Each reflection from the ionosphere is called a *hop* and allows radio waves to be received hundreds or thousands of miles away. This is the most common way for hams to make long-distance contacts. The highest frequency signal that can be reflected back to a point on the earth is the *Maximum Useable Frequency* (MUF) between the transmitter and receiver. The lowest frequency that can travel between those points without being absorbed is the *Lowest Useable Frequency* (LUF). This is illustrated in **Figure 2-36**.

When sky-wave propagation on an amateur band is possible between two points, the band is said to be *open*. If not, the band is *closed*. Because the ionosphere depends on solar radiation to form, areas in daylight have a different ionosphere above them than do those in nighttime areas. That means radio propagation may be supported in some directions, but not others, opening and closing to different locations as the earth rotates and the seasons change. This makes pursuing long-distance contacts very interesting!

Before you go on, study test questions T9B01, 02, 04-06, 10-11. Review this section if you have difficulty.

VHF and UHF enthusiasts also experience exciting ionospheric propagation. When solar radiation becomes sufficiently intense, such as during the peak of the 11-year sunspot cycle, the F layers of the ionosphere can bend even VHF signals back to earth. When those ham bands open, they support long-distance communication not possible under usual conditions. In addition, at all points in the solar cycle, patches of the ionosphere's E-layer can become sufficiently ionized to reflect VHF and UHF signals back to earth. This is called *sporadic-E* or E_s propagation and it is most common during early summer and mid-winter months.

Chapter 3

Operating Station Equipment

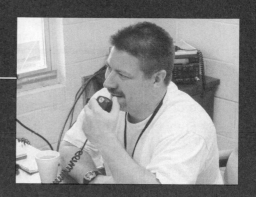

In this chapter, you'll learn about:

- **Basic operation of transmitters and receivers**
- **How repeater operation works**
- **Data mode basics**
- **Practical antennas and feedlines**
- **Power sources**
- **Simple test equipment for hams**
- **Special features of handheld transceivers**
- **RF Interference—Symptoms and Cures**

3.1 Transmitters and Receivers

In the first two chapters you have become acquainted with the basic equipment used by hams and the fundamental electronics and radio ideas that make the equipment go. We are now ready to start learning about Real Ham Radio, where knobs and dials get turned, meters jump, and signals crackle back and forth over the airwaves!

By the end of this chapter, you'll know your way around common radio controls and their functions. Figures and photos will show the typical controls used for these functions. Your radio will be different, of course, so rely on the operating manual for complete information.

You'll get a look at how antennas, feedlines, and power supplies are used by the typical amateur. We'll also cover the basics of dealing with interference. Knowing these practical techniques will get you started on the right foot as a ham. It also simplifies learning about operating conventions, rules, and regulations that follow. Ready? Let's find that power switch and get to it!

BASIC OPERATION—BAND, FREQUENCY AND MODE

Regardless of whether you use a transceiver (most likely) or a separate receiver and transmitter, there are two functions that are the same for all three—control of frequency and mode. Amateur radio is somewhat unique among the other radio services. Amateurs can tune anywhere in their assigned bands and are not required to use channels pre-assigned by the FCC, except in the 60-meter band. Repeaters operate on fixed channels, but hams developed that system on their own. Amateurs can also use many different modes, a few of which we covered in the previous chapter. Most other radio services are restricted to a single mode. **Figure 3-1** shows the typical frequency and mode controls of two common transceivers. If the radio is

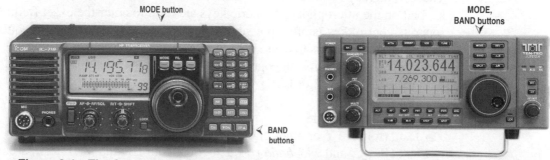

Figure 3-1—The frequency and mode controls are shown for a pair of simple HF radios, the ICOM IC-718 (left) and the Ten-Tec Jupiter (right). Frequency within can be adjusted with the large tuning knob or by the numeric keypad (IC-718). If a computer is connected to the radio, frequency may be changed by the computer software. Pressing the MODE key switches between CW, USB, and LSB.

Table 3-1

Band Selection Labels

Type of Labeling	Band Designations
By Frequency (MHz or GHz)	HF: 1.8, 3.5, 5, 7, 10, 14, 18, 21, 24, 28 MHz VHF/UHF: 50, 144, 220, 440, 902, 1.2 (GHz) or 1200 (MHz), 2.4 (GHz) or 2400 (MHz)
By Wavelength (meters or cm)	HF: 160, 80, 60, 40, 30, 20, 17, 15, 12, 10 (meters) VHF/UHF: 6, 2, 1¼ or 1.25 (meters), 70, 32, 23, 13 (cm)

a transceiver, frequency and mode settings apply to both transmit and receive.

With the Amateur Service's flexibility and freedom comes the obligation to know how to control frequency and mode. It is the operator that controls the frequency of the signal generated by the transmitter and acquired by the receiver. If the radio works on more than one band, such as a *dual-band* handheld transceiver for the 2-meter and 70-cm bands or a *multi-band* HF rig that covers 160 through 10-meters, the band is selected first. The band selection controls may be labeled in terms of frequency or wavelength. **Table 3-1** shows how the labels on your radio may indicate the band.

Then a frequency is selected within the band. The control used for continuous frequency adjustment is called a *VFO* for *Variable Frequency Oscillator*. In older radios, the VFO knob changed the value of a capacitor or inductor to change the frequency of an *oscillator*, a circuit that generates a steady RF signal. That signal determines the frequency on which the transmitter or receiver operates. In modern radios, the VFO control is used by a micro-processor to set the radio's frequency.

Many radios also have the ability to change the *tuning rate*, the amount of change in frequency with each turn of the VFO control. It's common to have a FAST button that causes the tuning knob to change frequency quickly. The STEP button or menu item, if available, gives the operator several choices of tuning rate by altering the *step size*; the smallest change in frequency caused by turning the VFO knob. A fast tuning rate works best for tuning in phone signals, while a slow tuning rate is used for CW and digital operating.

It is also common for radios to have multiple VFOs, usually a pair labeled VFO A and VFO B. In older radios, this meant two physical VFO circuits. Today, these VFOs are really just sets of data stored by the radio's microprocessor. Read your radio's operating or user manual for a complete description of VFO functions.

Some radios also have a keypad that the operator can use to enter frequencies directly. The keypad may be on the radio's front panel. Portable or mobile radios may place the keypad on the microphone. Numeric keys are used to enter the exact frequency, although a separate control key is often pressed first to let the radio know that a frequency is to be entered and afterwards to cause the radio to accept the frequency.

On *multi-mode* radios, the operator also selects the signal mode:
- AM or SSB (USB or LSB)
- FM
- CW or Morse code
- Data (for RTTY or other digital modes)

Most HF rigs and *wideband rigs* that cover both HF and VHF/UHF are multi-mode. Most handheld and mobile transceivers designed to be used with repeaters are single-mode radios that use only FM.

Memories or *memory channels* are used to store frequencies and modes for later recall. Memories are provided so that you can quickly tune to frequently used or favorite frequen-

With all the amazing functions and flexibility of today's radios, manufacturers have had to strike a balance between having separate controls for each function and having a reasonable number of buttons and keys. This means that many of the keys on your radio do double or even triple duty! If you look closely, you may see several labels in different colors or styles on a single key, each indicating a separate controlling action when the key is pressed.

To distinguish between the different actions, control or function keys are used, such as the MON-F key shown on the VX-7R handheld radio in **Figure 3-2**. (F stands for 'function' referring to the key's alternate use). When the MON-F key is pressed, the numeric keys all take on different meanings. For example, pressing the MON-F key changes the key usually reserved for the number 0 to a key that activates the receiver squelch setting function.

Some radios display menus of choices on their screen. The menu system is often activated with a MENU key. Inside the menu system, the radio keys have new functions to navigate the menu and select or adjust the various parameters and settings. Both keys and menus can be confusing, but reading the operator manual and a little practice will soon make it second-nature to use the most common functions, like setting your watch.

Figure 3-2—Because front-panel space is limited, some buttons and keys can perform more than one function. A control or function key causes the alternate functions to be used. On the VX-7R handheld a single MON-F key controls both alternate functions and writing information into the radio's memories.

cies. Depending on the type of radio, the memory channels may also store information such as power level and repeater access tones. Memories on VHF/UHF radios are often labeled as channels with each channel storing the specific pair of frequencies for use with repeaters. (Repeater operation is described later in this chapter.) With dozens of memories available on modern radios, the frequencies or repeaters you use are easy to access.

Other common keys for changing frequency or memory channels include UP/DOWN, MHZ, and BAND. These keys are found on the radio front panel or on the microphone. These make changing frequencies easy while mobile or when using the VFO would be inconvenient.

TRANSMITTER FUNCTIONS

In radios designed for AM/SSB/CW/Data transmissions, such as most HF and some VHF transmitters, the maximum transmitter output power is controlled by an RF Power control. A typical RF power control, usually a knob, is shown in **Figure 3-3**. (This may be a menu option or controlled by a knob shared by several features.) FM transmitters, such as handheld and mobile radios, generally have several fixed power levels selected by the operator. It is becoming more common to set power level as a menu option, since once set it's not often changed.

A second control that affects transmitter output power on AM and SSB transmitters is *Microphone Gain* (Mic Gain). This control adjusts the sensitivity of the transmitter's modulator circuits to the speech signal from the microphone. For AM and SSB transmitters, more modulation generally

Figure 3-3—The maximum output power of HF transceivers is set by an RF PWR knob. This control does not affect MIC GAIN or other audio settings. On mobile rigs, output power is varied between several preset levels with a button or key.

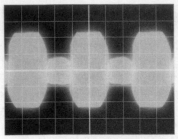

Figure 3-4—The MIC GAIN settings can have a big effect on your signal. If the MIC GAIN is too low, the signal is undermodulated and will sound weak. Too much MIC GAIN results in the distorted signal shown here. Notice that the peaks are flattened or rounded. The signal will likely cause interference to adjacent frequencies.

means more power in the output signal sidebands and a louder received signal. There can also be too much modulation (**Figure 3-4**), which is never a good thing. For FM transmitters, the output power doesn't change with higher modulation, but the signal's frequency deviation increases. Mic Gain has no effect on CW transmissions.

Most radios also have a *Speech Compressor* or *Speech Processor* control. Used on SSB, these circuits increase the average power of the transmitted signal by amplifying the weak parts of the speech signal more than the strong. While this creates a small amount of distortion, it improves reception when conditions are noisy or the received signal is weak. Compressors and processors should be used with care. Too much *compression* or *processing* creates severe distortion and can create interference on nearby frequencies.

Peak Envelope Power (PEP) is the measure of an AM or SSB signal's power. PEP is measured at the instant when speech into the microphone is the loudest. CW and FM transmissions have a constant power output, so PEP is measured as that constant level.

To avoid interfering with other stations while you're adjusting your transmitter or measuring its output power, it's a good idea to use a *dummy load*. A dummy load is a heavy-duty resistor that can absorb and dissipate the output power from a transmitter.

How do you get the transmitter to put out a signal, anyway? It's not enough to turn the transmitter on, you have to *key* it, as hams say, whether to transmit voice, CW, or digital data. There are two ways to key a rig. The first is *Push-to-Talk*, abbreviated PTT. The operator literally pushes the PTT switch on the microphone, closing a pair of electrical contacts that activate the transmitter. Once the transmitter is keyed, anything spoken into the microphone, keyed by a Morse key, or generated by a computer gets sent out over the air. Your rig will probably light up a "TRANSMIT" indicator, too.

The PTT switch may be on the microphone or in a footswitch. Most base and mobile radios also have a PTT input on an accessory connector so that a computer or data interface can key the transmitter.

While PTT works fine, using it for long periods of time can sure wear out your thumb. It also requires one hand to hold the microphone. The solution is *VOX*, which stands for *Voice-Operated Transmission*. VOX is a circuit that replaces the PTT switch, keying the transmitter whenever the operator is speaking into the microphone. Activating the VOX circuit is called *tripping the VOX*.

Since lots of sounds can enter the microphone—sneezes, wind, fans, and audio from the receiver—the VOX circuit has some adjustments of its own.

- *VOX Gain* controls the sensitivity of the VOX circuit to audio from the microphone.
- *VOX Delay* controls the amount of time the VOX circuit waits after you stop speaking before unkeying the transmitter.
- *Anti-VOX* prevents the speaker output from turning on the VOX circuit.

Splatter, Overdeviation, Clicks

Excessive modulation for all types of speech results in distortion of the received speech and unwanted or *spurious* transmitter outputs on adjacent frequencies where they cause interference. Those unwanted transmitter outputs have lots of names, but the most common is *splatter*. Generating those outputs is called *splattering*, as in, "You're splattering 10 kHz away!"

An overmodulated FM signal has excessive deviation and is said to be *over-deviating*. Over-deviation is usually caused by speaking too loudly into the microphone. An FM transmitter can also be misadjusted internally to over-deviate at normal speech levels. To reduce over-deviation, reduce your speaking volume.

Over-modulation of an AM or SSB signal as shown in Figure 3-4 is caused by speaking too loudly or by setting the microphone gain or speech compression too high. To eliminate over-modulation, speak more softly or reduce microphone gain or speech compression.

To help prevent overmodulating, *Automatic Level Control* (ALC) is a circuit that reduces output power as the transmitter's limit is reached. ALC is not a foolproof remedy, but will help keep your signal *clean*, that is, free of spurious outputs. Your radio's operating manual will show you how to use the radio's ALC meter to avoid over modulating your signal.

A signal that sounds overmodulated could also be the victim of RF energy. In other words, energy from your own transceiver could be getting into the microphone circuit and causing feedback.

Code transmissions can cause interference, too, by generating *key clicks*. Key clicks are sharp transient clicking sounds heard on frequencies near the offending signal. They are generated by a transmitter turning its output signal on and off too rapidly as it forms the dots and dashes. Key clicks can often be reduced by adjusting the transmitter's configuration settings for how it generates CW signals.

(A)

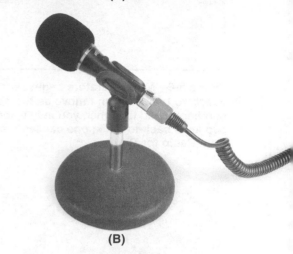

(B)

Microphones and Keys

The choice of microphone or key is very personal. When you buy your radio, a perfectly good microphone will usually be included. As you become experienced, you'll develop preferences for a certain style of operating and that includes the microphone and key. **Figures 3-5** and **3-6** show just a few of the many styles of microphones and keys.

Hand mikes, held in the hand during use, allow a handheld radio to be used while clipped to a belt or pocket. This keeps the antenna and the radiated signal away from your face and head—convenient and safe. Hand mikes may also feature operating and tuning controls. A speaker and microphone are often combined into one unit called a *speaker-mike*. *Desk mikes* sit on the operating table so that the operator does not have to hold them.

For use with VOX, *headsets* or *boomsets* combine a microphone with headphones so that the operator can use the transmitter with hands free. A *boom mike* is held in front of the face by a thin support (the *boom*) attached to the headphones. Boom mikes are particularly useful during contests or other long periods of operating. Boom mikes make it easy to use VOX, but many operators use a footswitch to control the PTT function instead.

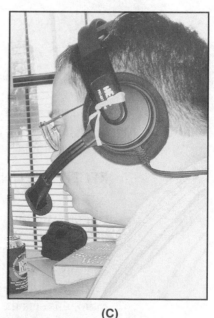

(C)

Figure 3-5—Several common types of microphones are shown in the photos. A hand mike shown in (a) is usually provided with the radio (in this case an FT-2800 by Yaesu). Desk mikes shown in (b) are popular at home stations with operators that like to have extended conversations, or ragchews. Contest and emergency communications operators prefer headsets as shown in (c) because they allow hands-free operation

Figure 3-6—Most operators begin sending Morse with a straight key (left). The key is attached to a table or board so that it doesn't move as the operator "pounds brass." When you've become more skilled and your sending speed is higher, you may want to switch to a paddle that connects to a keyer (right). The paddle is like two keys back-to-back; one causes the keyer to send dots and the other dashes. Courtesy of Wiley Publishing, Ham Radio for Dummies®.

Before you go on, study test questions T5A03; T5B01, 02, 03, 05, 07, 10, 11; T5D11; T9A07. Review this section if you have difficulty.

For sending Morse code, a *key* is used to turn the transmitter output signal on and off. Morse code's dots and dashes are known as the *elements* of the code. When using a manual or *straight* key, the operator generates the dots and dashes. This is called *hand keying*. The J-38 is a common type of straight key and is popular with beginners.

Once you are skilled at "the code", you'll want to go faster by using a *keyer*. This electronic device turns contact closures from a Morse *paddle* into a stream of Morse code elements. A keyer may be a standalone device or it can be built in to a transceiver.

A *paddle* is a pair of levers mounted side by side, each having its own set of contacts, one for dots and one for dashes. Keyers and paddles can generate Morse code much faster than by using a straight key.

RECEIVER FUNCTIONS

The receiver has a difficult task—picking just the one signal you want out of all the other signals on the airwaves. Nevertheless, modern receivers do a great job as long as they're adjusted properly. Knowing how a receiver's controls work makes a big difference, as you will see after you're licensed! **Figure 3-7** shows where you might find receiver controls on a typical receiver.

The most familiar control is the *AF Gain* or *Volume* control. Just like on a home or car radio, this sets the speaker or headphone listening level. On an HF rig, the *RF Gain* (Radio Frequency) control will be nearby. It adjusts the sensitivity of the receiver to incoming signals. *Attenuators* are used to reduce the strength of signals at the receiver input. They are used when excessively strong signals overload the receiver. Using the RF Gain and attenuator controls can remove a lot of distortion when the band is full of strong signals—give them a try!

While adjusting the AF Gain control always makes the output louder or softer, you'll notice that the RF Gain control doesn't have quite the same effect. That's because of the receiver's AGC circuitry. The *Automatic Gain Control* circuit constantly adjusts the receiver's sensitivity to keep the output volume relatively equal for both weak and strong signals. AGC can respond quickly or slowly to changes in signal strength. A *fast AGC*

Figure 3-7—The front panel of this Yaesu FT-840 transceiver has most of the receiver controls you'll use every day. AF GAIN controls volume. CLAR and SHIFT vary the receiver frequency without changing your transmit frequency. ATT adds attenuation to reduce very strong signals.

ATT

CLAR

SHIFT

response is used for CW and Data. *Slow AGC* response is used for AM and SSB signals. FM receivers don't use AGC because of the way they handle FM signals.

To keep from having to listen to continuous noise when no signal is present, the *squelch* circuit was invented. The squelch circuit mutes the audio output from the receiver when no signal is present. The *squelch threshold*, adjusted by the squelch control, is the signal level at which muting is turned off and the signal becomes audible. If the receiver's output is not muted, squelch is *open*. If the signal is muted, squelch is *closed*. Raising the squelch threshold is called *tightening the squelch*.

A receiver rejects unwanted signals through the use of *band-pass filters*. Every receiver passes radio signals through a sequence of increasingly narrow filters. Right at the receiver input, the band-pass filter may pass signals from an entire amateur band. Further into the receiver, in the circuitry known as the *IF* or *Intermediate Frequency* amplifiers, *IF filters* become narrow enough to reject signals that are adjacent to the one desired. IF filters are specified by the bandwidth of signals in Hz or kHz they pass. Wide filters (1.5 kHz or more) are used for voice mode reception. Narrow filters (1 kHz or less) are used for Morse code and data mode reception

Receivers have other weapons against noise and interference. A *notch filter* removes a very narrow range of frequencies from a receiver's audio output. This is useful when an interfering tone or whistle is encountered. The frequencies closest to the tone are removed, leaving the desired speech, code or data relatively unaffected. *Receiver Incremental Tuning* (RIT) is a fine-tuning control that allows the operator to adjust the receiver frequency without changing the transmitter frequency. This allows reception of a station that is not quite on the receiving station's frequency, or partially covered by noise, or to avoid interference.

There are lots of types of noise in the radio spectrum, too. Special circuits are employed to get rid of noises or at least cut them down to size. A *noise blanker* senses the sharp, buzzing pulses from arcing power lines, motors, or vehicle ignition systems and temporarily mutes the receiver during the pulse. A *noise limiter* doesn't try to remove the noise, it just prevents blasting your ears by *clamping* the audio signal at a controllable level. This is particularly useful when storms are nearby, generating powerful static crashes on the HF bands.

Receivers usually have a way to indicate signal strength. On a handheld transceiver, it is usually a variable-length bar at the side or bottom of the display. On the rig in **Figure 3-7**, the meter shows signal strength in *S-units*. Although not strictly calibrated, a change of one S-unit corresponds to about a factor of four change in signal strength. S-units are numbered from S1 to S9, with S9 being the strongest. The strength of signals stronger than S9 is reported as the number of dB greater than S9. i.e. - "Your signal is 20 dB over S9!"

Receivers frequently offer automatic tuning features called *sweep* and *scanning*. Sweep

Digital Signal Processing (DSP) —Radios with Software

DSP is the technique of using a special-purpose microprocessor in a radio to perform filtering and other functions on the received signal. Common DSP functions include:

• Noise Reduction
• Variable signal filtering
• Automatic Notch filtering
• Audio response tailoring on both receive and transmit

DSP works by converting the received signal into digital form. A microprocessor than performs the filtering and other functions on the digital data. Finally, the digitally modified data is converted back into an IF or audio signal and output through the speaker or headphones. DSP is making rapid advances and is a key part of many recent radios.

is smoothly tuning from one preset frequency to another. When *scanning*, the receiver rapidly jumps from preset channel to channel. The channels don't have to be adjacent or even in the same frequency band. Both sweep and scanning allow a receiver to monitor a band for activity. When the receiver encounters a signal, it will stop so the operator (that's you) can listen in. Amateurs frequently use scanning to listen to or *monitor* several different favorite repeaters without having to continually retune the radio. A radio made especially for scanning is called a *scanner*. Most commercial scanners can receive signals on one or more ham bands.

Beyond the Ham Bands

A receiver that can only be tuned to the bands used by amateurs is called an *amateur-band* or *ham-band* receiver. Of course, there are a lot more frequencies to tune than just the amateur bands. A receiver that can be tuned over that wider range is called a *general-coverage* or *wide-band* receiver. There are several types of such receivers:

• Short-wave—from several hundred kHz to 30 MHz
• Wide-band—from 30 MHz to 1.3 GHz or more
• Extended-coverage—VHF public safety and land mobile 174 MHz or UHF public safety and land mobile above 450 MHz.
• FM Broadcast—FM broadcast channels from 88 to 108 MHz.

Check the laws in your state regarding the legality of operating portable receivers or scanners that can receive law enforcement channels. Receiving mobile phone communications, paging transmissions, and encrypted communications is illegal.

> Before you go on, study test questions T5B04, 06, 09. Review this section if you have difficulty.

USING A REPEATER

A great deal of ham-to-ham communication is *simplex*. That means transmitting and receiving on a single frequency. Contact range and quality depend solely on the equipment of each station. If both parties can hear the other, all is well. Contacts can be difficult if one or both hams are operating while mobile or use a low-power handheld transceiver. This is where the helping hand from a repeater comes in.

As you learned earlier, repeater stations are located in high spots—on towers, hills, or buildings. The output of the repeater's receiver is retransmitted at the same time on a different frequency. The strong output signal is easily received over a wide area by mobile and handheld radios. Transmitting on one frequency and receiving on another frequency is called *duplex* communications.

In order to use a repeater, you will have to know three pieces of information; the repeater transmitter's *output frequency*, the repeater receiver's *input frequency*, and the frequency of any access control tones.

Repeater Shift

Let's start with the repeater's output frequency. This is the frequency on which you hear the repeater's transmitted signal. This is the frequency by which repeaters are listed in a directory, such as the *ARRL Repeater Directory*. Hams will say, "Meet you on the 443.50 machine" or "Let's move to the 94 repeater." (*Machine* is slang for repeater.) "94" means 146.94 MHz, a standard repeater channel frequency. To listen to

Table 3-2
Standard Repeater Offsets by Band

Band	Offset
10 Meters	−100 kHz
6 Meters	Varies by region: −500 kHz, −1 MHz, −1.7 MHz
2 Meters	+ or -600 kHz
1.25 Meters	−1.6 MHz
70 cm	+ or -5 MHz
902 MHz	12 MHz
1296 MHz	12 MHz

the repeater, tune to its output frequency.

To send a signal through the repeater, you must transmit on the repeater's input frequency where the repeater receiver listens. It would be chaos if every repeater owner used a different separation of input and output frequencies, so hams have decided on a standard separation between input and output frequencies. The difference between repeater input and output frequencies is called the repeater's *offset* or *shift*. The shift is the same for almost all repeaters on one band as shown in **Table 3-2**. If the repeater's input frequency is higher than the output frequency, that is a *positive shift*. *Negative shifts* place the repeater's input frequency below the output frequency.

Instead of having to remember two frequencies, using a standard shift allows you to remember only the repeater output frequency. The SHIFT or OFFSET key or menu setting on your radio allows you to switch between positive and negative shifts, or even no shift (simplex). Your radio is probably already configured to use the standard shift on each band. The operating manual has complete instructions on changing shift.

Repeater Access Tones

Most repeaters won't pass a signal from the receiver to the transmitter for retransmission unless it contains a special tone. The tone is one of 38 different frequencies, all below 300 Hz. Each repeater can have a different tone. The tone indicates to the repeater that your signal is intended for it and should be retransmitted.

Repeater access tones were invented by Motorola to allow different commercial users to share a repeater without having to listen to each other's conversations, these tones are known by various names; *Continuous Tone Coded Squelch System* (CTCSS), *PL* (for Private Line®, the Motorola trade name), or *subaudible*. FRS/GMRS radio users know these tones as *privacy codes* or *privacy tones*.

Why wouldn't a signal on the proper frequency be intended for the repeater? Most repeaters are installed close to other repeaters, paging transmitters, and radio and television transmitters. The powerful signals from all these transmitters sometimes mix together and create false signals, called *intermod*, which is an abbreviation of *intermodulation*. Intermod can easily appear on a repeater's input frequency and would be retransmitted, annoying listeners since the audio portion is usually garbled or data bursts, disrupting normal communications. To prevent these signals from being retransmitted, the repeater receiver listens for the proper tone in the received signal. No tone or an improper tone usually means the signal is not intended for that repeater.

Your radio will have a TONE key or menu selection that allows you to both select a tone and add it to the transmitted signal, if desired. You may also be able to set your radio's squelch to require a CTCSS tone to pass received audio to the speaker. This is called *tone squelch*. You should be aware that most repeaters filter out CTCSS tones before the received audio is retransmitted.

A third method of squelch control is *Digitally Coded Squelch* (DCS). A short burst of tones is sent with your transmission. If the proper tone sequence is received, your receiver will open up the squelch and you can hear the calling station. Check the operating manual of your radio for information about how to configure DCS.

Before you go on, study test questions T5B08; T5C03, 05-08 Review this section if you have difficulty.

DIGITAL DATA MODES

As you learned in the previous chapter, digital data modes combine modulation (the addition of information to a radio signal) with a protocol (the rules by which data is formatted and packaged). To use a data mode, you'll need a data interface that make the connection between the radio and whatever external equipment is used to read and generate data. For example, to use packet radio you'll need a TNC or *Terminal Node Controller*. If you use several digital modes, you may want to use an MPC or *Multiple Protocol Controller*.

For many protocols, the conversion between digital data and audio tones can be done by a combination of software and the sound card in your computer. You'll still need a data interface between the computer and the radio. Data interfaces electrically *isolate* the radio and computer so that unwanted effects of hum or RF feedback cannot interfere with the data signals. **Figure 3-8** shows a generic example of how a station's data interface is connected.

To make sure your digital signal is transmitted and received correctly, you'll need to adjust or configure all of the following:

- Transmit audio level—all of the same cautions apply as with voice operation about over-modulation due to excessive input signal strength to the transmitter.
- Receive audio level—the output from the receiver must be at the proper loud ness or level for the data interface to turn it back into data. Levels that are too high distort the data tones and if too low, allow too much noise to be mixed in. Both situations cause errors in the tone-to-data conversion.
- Digital interface—if you are using a computer, you may need to configure the connection to the data interface so that the proper control signals are connected.
- Transceiver control—turning the transmitter on and off at the right time may require a connection to the PTT (Push-to-Talk) input of the radio.

The operating manual supplied with your data interface or software will show you how

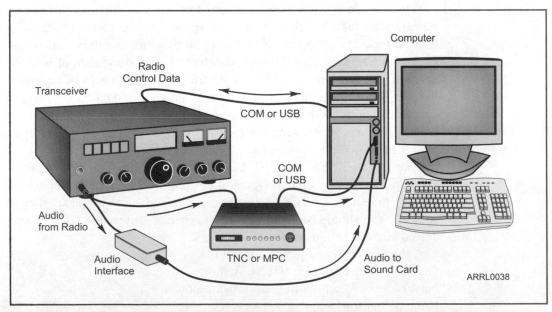

Figure 3-8—Data interfaces are connected between the transceiver's audio inputs and outputs and the computer's data connections (USB or COM ports) or sound card jacks. A TNC or MPC converts between data and audio. An audio interface only isolates the computer sound card from the radio to prevent hum. Courtesy Wiley Publishing, Two-Way Radios and Scanners for Dummies®.

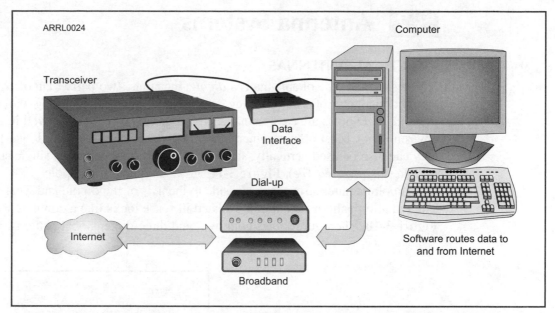

Figure 3-9—An Internet gateway station acts like a regular digital mode station except that the computer runs software that relays data between the radio and the Internet. The most common example of gateway stations is Winlink PMBO stations.

to make all the necessary adjustments on the radio or computer. Once the adjustments are set properly, record the position of your audio and microphone gain controls and save the computer sound card settings. You can then return to digital operation quickly.

A special kind of digital station provides a connection to the Internet for data transmitted by Amateur Radio. This kind of station is an *Internet gateway* as shown in **Figure 3-9**. Most gateways a set up to *forward* messages. The most common examples are packet radio *bulletin board systems* (BBS) and the Winlink system's *Participating Mailbox Operators* (PMBO). Messages with a recognized Internet email address can be sent and retrieved over these systems. Learn more about the Winlink system at **www.winlink.org**.

Before you go on, study test questions T5A08-10; T6A03. Review this section if you have difficulty.

Another type of gateway provides direct Internet connectivity so that a computer running standard Web browser software can connect to any Internet address. The radio has a built-in data interface with an Ethernet connection to which the computer connects, just like a home network. This type of gateway is provided by radios using the D-STAR system.

Caution! All of the rules and regulations about commercial and business-related messages and communications apply to Internet gateways. For example, it is definitely not okay to exchange e-mails for your employer or to access Web sites with third-party advertising. Because so much of the Internet is associated with commercial activity, take extra care to abide by the restrictions of Amateur operating rules.

3.2 Antenna Systems

PRACTICAL ANTENNAS

The simplest type of antenna is a *dipole*, meaning "two parts." Dipoles are made from a straight conductor of wire or tubing one-half wavelength ($\frac{1}{2}$-λ) long with a feedpoint in the middle. Dipoles are easy to make, easy to use, and work quite well in a variety of environments. Most often oriented horizontally, particularly on the lower frequency bands, they can also be used vertically, sloping, or even drooping from a single support in the middle (the *inverted-Vee*). **Figure 3-11** shows a typical wire dipole.

A dipole radiates strongest broadside to the axis of the dipole and weakest off the ends. The radiation pattern for a dipole isolated in space looks like a donut as seen in **Figure 3-10**. The figure shows both two- and three-dimensional patterns. The two-

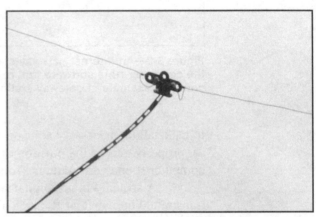

Figure 3-11—A basic dipole antenna.

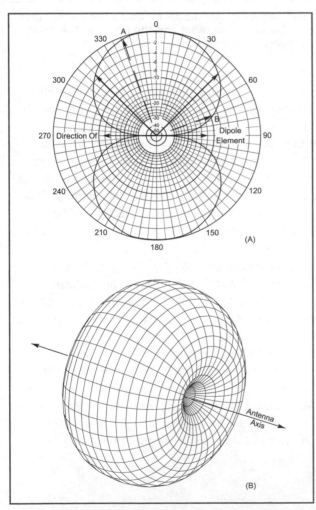

Figure 3-10—The radiation pattern of a dipole far from ground (in free-space). At (a) the pattern is shown in a plane containing the dipole. The lengths of the arrows indicate the relative strength of the radiated power in that direction. The dipole radiates best broadside to its length. At (b) the 3-D pattern shows radiated strength in all directions.

How Long Is That Dipole?

If a dipole is $\frac{1}{2}$-λ long, then how long is that in feet? Starting with the frequency at which the dipole is to be resonant, here's a handy formula:

Length (in feet) = 468 / Frequency (in MHz)

or l = 468/f

Example: A 6 meter band dipole for 50.1 MHz is 468/50.1 = 112 inches long.

This formula accounts for the effect of wire thickness on antennas which shortens antennas a few percent compared to a signal's wavelength traveling in air. Dipoles have to be shortened a few percent from the formula length. If you are making the dipole out of wire, be sure to cut a little extra wire so that it can be fastened to insulators and supports. Make the dipole a few percent long at first, then use your SWR meter to adjust it to the length that is resonant at your desired frequency. This allows you to compensate for the effects of ground or nearby conductors that might affect the antenna.

dimensional pattern shows what the three-dimensional pattern would look like if cut through the plane of the dipole.

Another popular antenna is the *ground-plane* antenna. The most common type of ground-plane antenna is one-half of a dipole (¹/₄-λ long) with the missing portion made up by an electrical mirror, the ground plane, made from sheet metal or a screen of wires called radials. The ground-plane is generally one-quarter wavelength long with the feedpoint at the base of the antenna, as shown in **Figure 3-12**. One conductor of the feedline is connected to the antenna and the other to the ground plane. The length of a ground-plane is half that of a dipole: length (in feet) = 234 / frequency (in MHz).

What is meant by "electrical mirror"? Have you ever watched as someone put their face at the end of a mirror, the reflection recreating the face's missing half? A ground-plane works the same way, as illustrated in Figure 3-12. The more complete the ground-plane plane, the better the reflection and more dipole-like the result. A good ground plane should extend at least ¼-λ from the base of the antenna in all directions. As you might expect, since a ground-plane antenna emulates a dipole, it radiates strongest perpendicular to the axis of the antenna and weakest off its end.

Ground-plane antennas are often called *verticals* because it is easy to mount them perpendicular to the ground, with the ground-plane on or parallel to the ground. When mounted this way, the radiation pattern of the ground-plane is *omnidirectional*, transmitting and receiving equally well in all directions. This type of pattern is used when there is no preferred direction of communications.

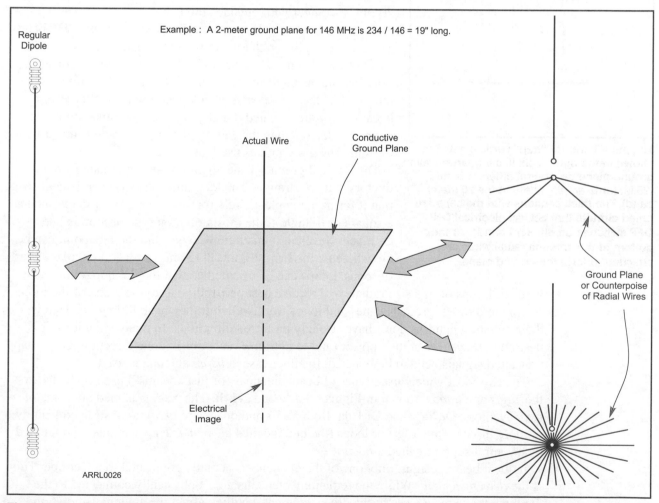

Figure 3-12—A ground-plane makes up an electrical mirror that creates an image of the missing half of a ground-plane antenna. The result is an antenna that acts very much like a dipole. The ground plane can be made up of a screen of wires (often used at HF) or for smaller VHF and UHF antennas mounted on masts, a counterpoise of a few wires serves the same purpose.

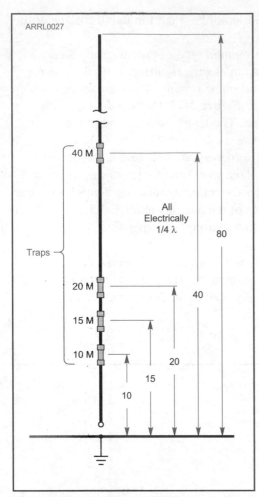

ARRL0027

40 M

All
Electrically
1/4 λ

80

Traps

20 M

40

15 M

10 M

20

15

10

The multi-band HF "trap" vertical antenna shown in the figure acts like a quarter-wave ground plane on several different bands (20 M, for example, refers to the 20 meter band). The thick sections with the caps are tuned circuits that act like electrical ON/ OFF switches on different bands. At the bottom of the antenna radial wires are attached to form the ground plane.

Good examples of practical vertical antennas are the small whips placed on the roofs or trunks of cars. The metallic body of the car makes up the ground plane. On the ham bands at 50 MHz and above, these antennas give great performance at low cost. Even better than the ¼-λ version, the 5/8-λ vertical, due to its extended length, focuses a bit more energy towards the horizon, improving range.

Whip antennas for VHF and UHF can be mounted on vehicles in many ways. One of the most popular methods (requiring no mounting holes) is the *mag-mount* or *magnet-mount* antenna. The mag-mount, has a large magnet in its base to hold it tightly to a steel surface. The center conductor of the coaxial feedline is connected to the antenna and the outer shield to a thin metal plate held next to the ground plane by the magnet. Even though the metal plate is not directly connected to the ground plane, it is so close that it acts like a capacitor, passing the ac current of the radio signal as if it were connected. (Note—mag-mount antennas won't work on a plastic car body and won't cling to aluminum or non-ferrous metals!)

An antenna on the outside of your car is vastly superior to a typical handheld "rubber duck" used inside. Signals can be 10 to 20 times stronger with outside antennas.

Ground planes are also used on the lower-frequency HF bands. Since wavelengths on these bands can be many meters, a solid sheet of metal is not often an option for ground plane construction. Instead, as indicated in Figure 3-12, a number of wires are laid out like the spokes of a wheel. These wires, called *radials* because they are oriented radially from the antenna, are much better conductors than the earth and so form a better mirror than the soil. The more radials used, the better.

Dipoles and ground-plane antennas are both made from conductors that are straight. The loop antenna, however, is bent so that it forms a complete circle (or any other closed shape such as a square or triangle). The most common type of loop is one complete wavelength in circumference and the feedpoint can be anywhere on the loop. **Figure 3-13** shows several popular styles of loops. Loops can be oriented so that they're horizontal or vertical. Loops of this size radiate and receive best perpendicular to the plane of the loop.

Simple dipoles, ground-planes and loops work well, but they have little *gain*. That is, their radiation patterns don't have strongly preferred directions. In many situations, it is desired to focus transmitted power (and to optimize reception) in one direction. Or perhaps unwanted signals need to be rejected. In either case, a *beam* antenna is used.

The two most widely used types of beam antennas (or just "beams") used by hams are the *Yagis* and *quads*, shown in **Figures 3-14a** and **3-14b**. The Yagi is named after one of its two inventors, Dr Yagi and Dr Uda. Both Yagis and quads are *arrays*, constructed from two or more dipoles (the Yagi) or loops (the quad) called *elements*. The elements are mounted on a central support called a *boom*.

In most beam designs, only one of the elements is actually connected to a feedline. This is the *driven element*. While the remaining elements aren't physically connected to the feedline, they are so close to the driven element that they affect the antenna's radiation pattern. These are called *parasitic elements* because they alter the radiation pattern of the driven element without being connected to it. The length of the parasitic elements and their arrangement along the boom cause the antenna to focus energy along the boom in

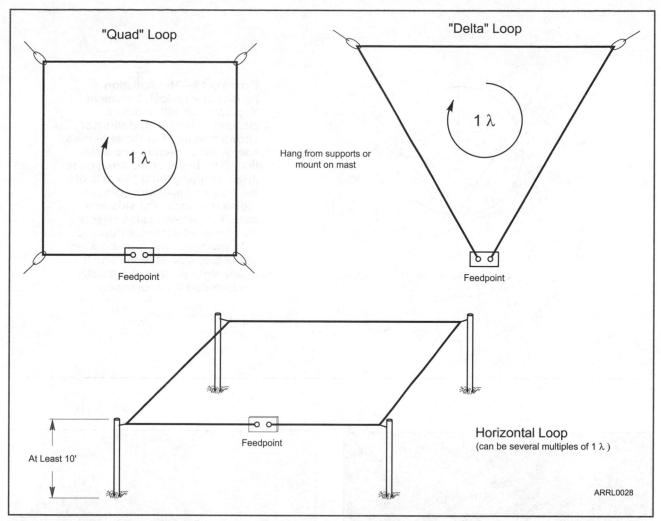

Figure 3-13—Most loop antennas are 1-wavelength in circumference and can be fed anywhere along the antenna. Quad and delta loops are small enough on the higher HF bands and at VHF and UHF that they can be rotated to cover different directions. A horizontal loop can be used on several HF bands and may be much larger than 1-wavelength.

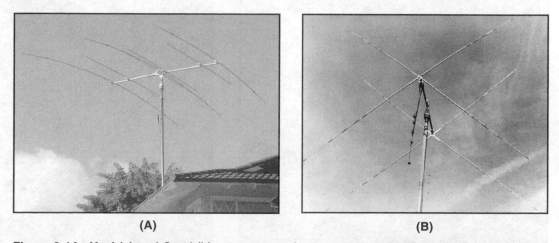

Figure 3-14—Yagi (a) and Quad (b) antennas are beam antennas that focus radiated power in one direction along their axis. Yagis use tubular elements and can be oriented horizontally (as shown here) or vertically. Quads use quad loops as the elements. By feeding the driven element at different points the resulting antenna can be either horizontally or vertically polarized.

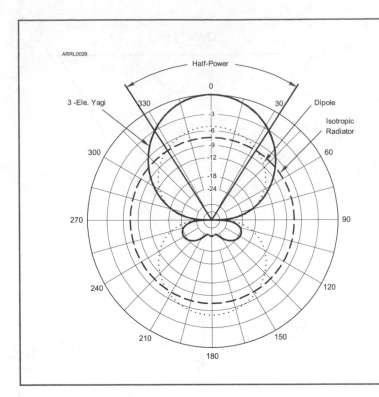

ARRL0029

Figure 3-15—The radiation pattern of a typical, 3-element Yagi antenna with a driven element, reflector, and director shows that most of the antenna's energy is focused in one direction. (The boom of the antenna is oriented along the 0-180 axis of the graph.) Small amounts are radiated towards the side and back. This antenna also rejects noise and interference from the side and back. The round pattern of the isotropic antenna and the figure-eight pattern of a dipole are included for reference.

(A)

(B)

Figure 3-16—At (a) N7AM's 10-meter diameter homebuilt dish is used on 1296 MHz for moonbounce (EME) and long-distance terrestrial contacts. N7CFO's much smaller dish at (b) operates on 10 GHz and is portable enough to be taken on contest outings.

one direction. The radiation pattern for a typical beam is shown in **Figure 3-15**.

Careful inspection of these antennas will show that parasitic elements located along the boom in the direction of maximum gain are slightly shorter than the driven element. These elements are called *directors*. Parasitic elements in the direction of minimum gain are called *reflectors* and are slightly longer than the driven element.

As frequency increases and the length of Yagi and quad elements becomes smaller, it becomes more difficult to construct practical antennas. At frequencies above 1 GHz, a different style of antenna becomes practical—the *dish*. Amateur dish antenna work very much like the satellite TV dishes springing up on homes and apartment railings. **Figure 3-16** shows a few examples of amateur dish antennas.

Before you go on, study test questions T9A01, 02, 05, 06, 08, 09, 10, 11, 12. Review this section if you have difficulty.

A dish antenna has much more gain than a Yagi—a factor of 10 or more! On the 33 cm and 23 cm bands, dishes of a few feet in diameter are common and really large dishes are not unknown. At higher frequencies of 10 GHz or more, dish size shrinks below ½-meter, while still providing plenty of gain.

PRACTICAL FEEDLINES AND ASSOCIATED EQUIPMENT

Feedlines and Connectors

By far, the most common type of feedline used to connect the antenna to the radio equipment is coaxial cable. You'll often recognize its black outer sheath, which provides protection from damage by ultraviolet light. The most popular types of "coax" used by amateurs are shown in **Table 3-3**. There are many variations on these basic types, but all have similar characteristics. After a characteristic impedance is chosen (usually 50 Ω), the most important consideration is cable loss—the amount of power lost in the form of heat as signals travel along the cable. Loss increases with length and frequency. Using coax for very long runs or very high frequencies may result in unacceptably high losses.

Loss is specified in dB per 100 feet of cable at a certain frequency. Table 3-3 gives cable loss at 30 MHz (close to the 10-meter band) and at 150 MHz (close to the 2-meter band). How much loss is too much? In general, losses of 3 dB or less are barely noticeable. Losses of 6 dB or more begin to make a serious dent in your contact range and quality—look for another solution! You can change to a lower-loss cable or reduce the cable length.

Open-wire lines have much lower loss than coaxial cable. For example, at 30 MHz, 450 Ω "window" line has a loss of only 0.15 dB/100 feet while RG-58 loses 2 dB in the same length. At 150 MHz, RG-58 is losing about 4.5 dB/100 feet and the window line a bit less than 0.4 dB/100 feet. Why don't hams use open-wire feedlines instead of coax? It's primarily an issue of convenience. Because all of the signal energy is contained

Table 3-3
Types of Coaxial Cable

Type	Impedance Ω	Loss per 100 feet (in dB) at 30 MHz	Loss per 100 feet (in dB) at 150 MHz
RG8U	50	1.8	6.9
RG8X	50	3.7	12.8
RG11U	75	3.2	11
RG58U	50	3.2	4
RG59U	75	1.6	3.6
RG174U	50	8.9	28.2
RG213U	50	2.2	8
9913	50	1	4.5

completely inside coax, it can be in almost any location. Most antennas are designed to have feedpoint impedances that are close to that of coaxial cables, as well. Open-wire, on the other hand, must be kept clear of metal, other conductors, and other cables. Its higher impedances don't match well with most transmitters.

When would you use open-wire line? If you needed a really long run of feedline it would be a good choice, if it could be kept clear of nearby conductors. Antennas that operate on many different frequencies often present extreme feedpoint impedances, creating a high SWR in the feedline. In coaxial cable, high SWR means that a lot of power is bouncing back and forth in the line, losing energy as heat with each trip. Open-wire's much lower loss enables operation on these frequencies without big feedline losses.

In order to prevent excessive losses, feedlines must be protected. Coaxial cable depends on the integrity of its outer coating—the jacket—to keep water out. Nicks, cuts, and scrapes can all breach the jacket. Water in coaxial cable degrades the effectiveness of the braided shield and dramatically increases losses. Coax cannot be bent sharply, lest the center conductor be forced gradually through the soft center insulation, eventually causing a short. Prolonged exposure to the ultraviolet (UV) in sunlight will also cause the plastic in the jacket to degrade, causing small cracks. Coax that has been exposed to the weather for a long time often has losses much higher than that of new cable.

Open-wire feedlines also need care. These feedlines often are constructed using solid wire. Prolonged flexing in the wind will eventually crack and break the conductors if no strain relief is provided. Moss, vines, or lichen growing on the cable will also increase loss. Tree limbs rubbing on the line will eventually break it. Protect splices from

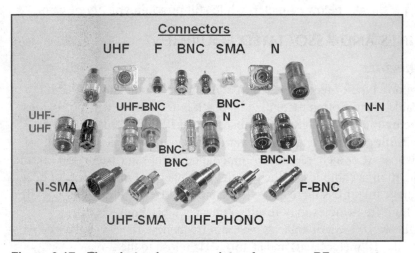

Figure 3-17—The photo shows a variety of common RF connectors that hams use. The larger connectors are used for higher power transmitters and antennas. The most common are the UHF and BNC styles. Special adaptors are used to make connections between cables and equipment that have different styles of connectors.

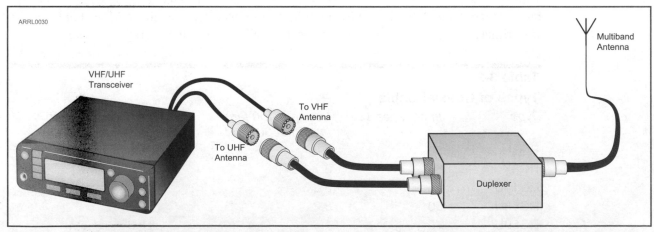

Figure 3-18—The duplexer splits RF signals coming from the antenna into high- and low-frequency paths. Low-frequency signals coming from the radio are routed to the antennas and not the high frequency connection or port. High-frequency signals are routed in a similar way. A triplexer has three radio ports instead of just two so that a single antenna can be used on three bands.

weather damage with good-quality electrical tape or a paint-on coating.

Connectors for coaxial cable ("coax connectors") are required to make connections to radios, accessory equipment, and most antennas. "Pigtail" style connections, where the braid and center conductor are separated and attached to screw terminals are generally unsuitable at frequencies above HF. Pigtails also expose the cable to water.

Figure 3-17 shows several common types of coaxial connectors. The figure also shows *adaptors* that make connections from one type of connector to another. Complete information on common coax connectors, including assembly instructions, is available in any edition of the *ARRL Handbook for Radio Communications*. Learning how to put on your own connectors not only saves money but also will allow you to make repairs at home and under emergency conditions!

Feedline Devices

Owners of multi-band VHF/UHF radios often want to use a single antenna on all bands. There are several models available that cover 50/144/440 MHz, for example. But the radio may have a separate output for each band. A *duplexer* is the solution. It allows the two radio outputs, operating on different bands, to share a single antenna. There are even *triplexers* for three-band antennas. **Figure 3-18** shows an illustration.

Another common feedline device is the *balun* (pronounced "băl-un"). Balun is an abbreviation of "balanced to unbalanced." The primary function of a balun is to keep current from flowing back down the outside of a coaxial cable shield at the connection to the antenna. Current flowing there, instead of on the antenna, may flow back into your "shack" where it

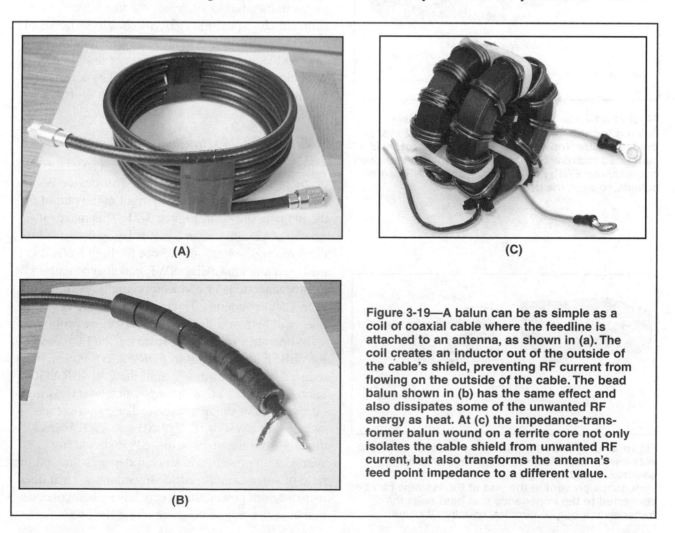

(A)

(C)

(B)

Figure 3-19—A balun can be as simple as a coil of coaxial cable where the feedline is attached to an antenna, as shown in (a). The coil creates an inductor out of the outside of the cable's shield, preventing RF current from flowing on the outside of the cable. The bead balun shown in (b) has the same effect and also dissipates some of the unwanted RF energy as heat. At (c) the impedance-transformer balun wound on a ferrite core not only isolates the cable shield from unwanted RF current, but also transforms the antenna's feed point impedance to a different value.

Figure 3-20—Coaxial or antenna switches are made to switch a single feedline between two or more antennas or pieces of equipment. They are specially made to handle a transmitter's full output power without disturbing the feedline's characteristic impedance.

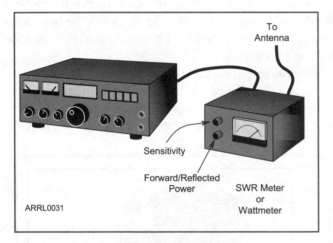

Figure 3-21—The SWR meter measures power flowing towards the antenna (forward) and towards the transmitter (reflected or reverse). The meter is calibrated to show SWR or power. A wattmeter does not measure SWR, just power, and the SWR can be calculated from the power readings.

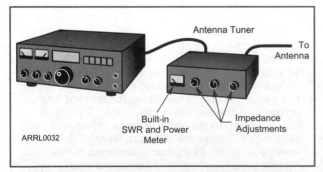

Figure 3-22—The antenna tuner shown in the photo acts like an electrical version of a mechanical gearbox. By adjusting the tuner's controls, the impedance present at the end of the feedline can be converted to the impedance that best suits the transceiver's output circuits, usually 50 ohms.

can cause RF feedback or interference with telephones and computers. This type of balun can be made from a coil of feedline on the HF bands or from ferrite beads (ferrite is a magnetic ceramic we'll learn about shortly) slipped over the outside of the cable. **Figure 3-19** shows some examples of baluns. These baluns are also called *choke baluns*, since they "choke off" RF current.

Another type of balun also transforms the impedance of an antenna to match coaxial feedlines. For example, a 9:1 balun would be used to transform an antenna feedpoint impedance of 450 Ω down to 50 Ω, where it matched the cable's impedance. This allows all of the RF power flowing up the feedline to be transferred to the antenna. These baluns are called *impedance-transformer baluns* and are available in a variety of ratios; 2:1, 4:1, 9:1, up to 16:1.

Finally, if you have one transmitter and several antennas that are designed for different bands, it would be inconvenient to disconnect and connect antennas as you change frequencies. An *antenna switch* is used to connect one of several antennas to a single transmitter. Switches come in a wide variety from simple 2-position A/B switches to switches that allow two transmitters to share several antennas. To save on feedline, *remote switches* are used that do the switching near the antennas, selected by a control box in the shack. A six-position switch is shown in **Figure 3-20**.

SWR Meters, Wattmeters and Tuners

To measure the SWR in a feedline a special type of meter, the *SWR meter*, is used. This device is placed in the feedline, usually right at the output of the radio as shown in **Figure 3-21**. This makes it easy to see exactly what SWR is being presented to the radio and to keep an eye out for high SWR due to mistuned antennas. High SWR can also be caused by loose connections or connectors, faulty feedline, or even a faulty antenna. Having an SWR meter "in the line" will help you discover and fix those problems.

To operate a manual SWR meter, the FORWARD-REVERSE switch is set to FORWARD. Power is reduced on the transmitter until the CALIBRATION knob can be set so that the meter needle travels to the far right end of the meter scale. The meter is then switched to SWR or REVERSE and SWR is read directly from the meter scale. SWR should not change with power level since it depends only on the ratio of antenna and feedline impedances. (If it does increase with power, you likely have a bad connection or connector somewhere along the line.)

Instead of SWR meters, many amateurs prefer a *wattmeter* and better yet, a *directional wattmeter*. Wattmeters measure power in a feedline and can be placed in the line to read power flowing in either direction. Directional wattmeters can measure power flowing towards the antenna and power reflected from the antenna by rotating a sensing element or turning a switch. The operator can then convert the forward and reflected power readings to SWR by using a table or formula. **Figure 3-21** shows examples of where SWR meters and wattmeters are connected.

If SWR at the end of the feedline is too high for the radio to operate properly, the devices in **Figure 3-22** called *impedance matchers* or *transmatches* or *antenna tuners* are used. These gadgets act like electrical gearboxes, transforming the impedance presented by the feedline to an impedance more suitable to the radio's transmitter output. This is just like a car's transmission, which allows the engine's high-speed rotations to supply power to the lower-speed wheels.

Before you go on, study questions T9C07, 09, 10.

Note that the antenna is not really tuned—it's the impedance at the output of the feedline that is adjusted. This allows the transmitter to output at full power output without the high SWR causing damage. For convenience, most tuners combine the functions of impedance matcher, directional wattmeter, and antenna switch. There are also *automatic tuners* that sense when SWR is high and make the necessary adjustments under the supervision of a microprocessor to match the feedline and transmitter impedances.

ANTENNA SUPPORTS

Towers and Masts

To increase contact range a tower or mast is used to raise antennas above buildings and other obstructions. This has the added benefit of reducing interference by signals to equipment, such as telephones or televisions. This works two ways, since consumer

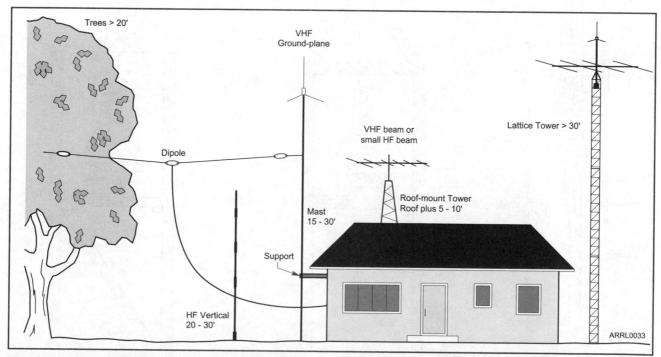

Figure 3-23—The drawing shows several common ways of holding up antennas. If you have trees, wire antennas such as dipoles are an easy and inexpensive option. Masts up to 30 feet or more can hold small ground-plane antennas. Rooftop towers are available in a number of sizes for beams. You can also go all the way and erect a lattice tower that can hold substantial antennas many feet in the air.

electronics often generate weakly interfering signals of their own! Distance makes good neighbors in the radio world. **Figure 3-23** illustrates several way of supporting antennas.

A fixed-length pipe mast is the simplest method of raising small antennas, such as ground planes or small directional antennas. This works well for heights up to 20' and is fairly inexpensive, requiring only the support of a building to keep the mast and antennas in the air. For higher reaches telescoping *push-up masts* for small antennas are available up to 40'. These are the same masts used by TV antenna installers for many years.

Above 40 feet a *lattice tower* is used. Lattice towers come in 8 or 10-foot sections and are used by amateurs at heights up to 200 feet. These towers are constructed with legs made of pipe and welded cross bracing. They are also required to support large antennas, such as HF Yagis and quads or large VHF/UHF antenna arrays. The tower is either supported by a building or is mounted on a concrete base and guyed.

If you don't care to assemble and climb a full-sized guyed tower, *crank-up* and *tilt-over* towers can support large antennas at heights of 70 feet or more. Crank-up and tilt-over towers are generally self-supporting, but require a substantial concrete base and must be lowered to avoid strong winds and storms.

Regardless of what type of antenna and supporting structure you choose, it's likely that if you live in a suburban or urban area, regulations will apply. Building permits are generally required for lattice, crank-up, and tilt-over towers. Homeowner's association rules or covenants may restrict your antenna choices. Be aware of these before you start your antenna projects.

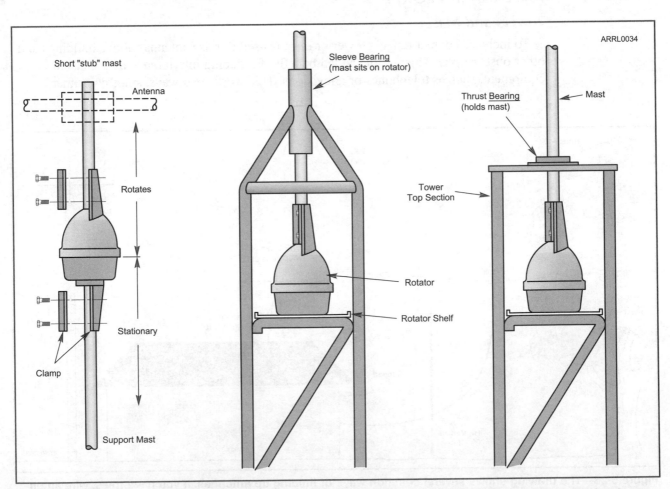

Figure 3-24—The stationary half of rotators are clamped or bolted to a fixed platform, such as a mast or tower. The moving part holds a mast on which the antenna is mounted. A control box in the radio room is used to point the antenna in different directions.

Rotators

Sometimes communications in only a single direction is needed. For example, if you lived far from a city and only needed to contact stations there. In that case, you could mount your antenna on a mast or tower fixed to point in one direction. More often, hams want to be able to contact stations in several or all directions, so it's necessary to be able to point the antenna in those directions as desired. To do that, a *rotator* is used.

A rotator (not a rotor, which is something that rotates) consists of a motor unit mounted under the antenna connected with a multi-conductor cable to a control unit near the operator. **Figure 3-24** shows the basic idea. The motor unit is contained in a strong waterproof housing with a clamp on the top for a pipe *mast* on which the antenna is mounted. If the rotator is mounted in a tower, its base is bolted to a *rotator shelf* at the top of the tower. Rotators can also be mounted on fixed masts with a clamp on the bottom of the motor unit. Only the top of the motor unit turns, while the base is firmly held to the unmoving support. The control unit has a direction indicator, usually a meter, and switches to turn the rotator in either direction.

Construction Materials

For antenna construction, copper wire or tubing and aluminum tubing or rod are usually selected. Wire antennas are most often constructed from copper or copperweld wire and ceramic or plastic insulators. Aluminum is most often used to construct beam antennas. Copper and aluminum are both excellent conductors. Aluminum is easy to work, strong, and light, as well, but is very difficult to solder.

Towers and masts are usually made from galvanized steel. There are aluminum towers available at somewhat higher expense. Galvanizing prevents rust and, if maintained, will last for many years. Galvanized bolts and nuts are also available for assembly. If wood is used as a support, galvanized lag and carriage bolts are recommended.

To assemble aluminum antennas and other outdoor projects, stainless steel hardware resists corrosion and will last for many years. Avoid nickel or cadmium-plated hardware as it will rust quickly, making maintenance or repairs much more difficult.

Ropes and halyards should be UV-resistant to avoid degrading in direct sunlight. Synthetic rope and cord is available with special fillers and pigments that block the harmful effects of UV. Avoid yellow "poly" ropes and cotton or fiber ropes as they will not hold up, letting your precious skyhooks tumble back to earth!

Waterproofing is also important, particularly coaxial connectors and other electrical connections. Connectors should be wrapped in at least two layers of high-quality electrical tape, such as Scotch® 33 or 88. Apply a small amount of waterproofing compound such as Oxygard™ to exposed screw terminals and cover them, if possible. Water and electricity don't mix with good results, whether at 60 Hz for power or in the MHz for radio!

Before you go on, study test question T0B10. Review this section if you have difficulty.

3.3 Power Supplies and Batteries

The most exciting part of radio is, of course, radio signals and power, but a solid power source is just as important. You'll find that a healthy source of power makes for a clean, noise-free signal and better reception. Using a power source whose output voltage is too high or too low or that cannot supply sufficient current can damage radio equipment or make it operate improperly.

Power Supplies

Power supplies that operate from household ac power are the most common source of power for radios. One is shown in **Figure 3-25**. They convert the ac input power to a smooth dc current that the radio can use. A power supply has two primary ratings; the amount of current it can supply continuously at a specific voltage, such as a "12 V, 20 Amp" supply. The output voltage may vary or be adjustable, so the voltage rating is a *nominal* rating. For example, the "12 Volt" supply's output might be anywhere between 11.75 and 12.25 volts.

A supply's output voltage changes with the amount of output current. The percentage of voltage change between zero current (*no load*) and maximum current (*full load*) is the *regulation* of the supply. To achieve "tight" regulation, meaning little variation as current changes, requires a *regulator* circuit in the supply. These supplies are *regulated supplies* and have regulation of a few percent, the output voltage varying only a small amount with load. Regulated supplies are the best power source for radios.

The current rating of a supply must be at least as much as the sum of the maximum current needs of everything hooked up to the supply. The manual for each piece equipment should tell you how much current is required. A single power supply can be shared between two or more pieces of equipment if it can supply enough current.

Figure 3-25—The traditional or "linear" power supply is a reliable, if heavy and bulky, performer. The modern "switching" supply shown here uses sophisticated electronics to avoid large iron transformers, delivering equivalent power at a fraction of the weight.

Generators & Inverters

For portable and emergency operation, ac power from the utility grid might not be available. In situations like this, generators and inverters are used. A *generator* turns energy from an engine into ac or dc power. An *inverter* turns dc power into ac power. The ac power from either can then be used to run a regular ac power supply or charge batteries.

Generators and inverters are rated in watts, the amount of power they can supply while still maintaining output voltage within the specified range. Rarely can either be used at full power continuously, so it is best to be conservative when choosing or using them.

Electronic equipment that runs from ac power generally expects its input current to be a relatively undistorted sine wave. A distorted output from an inverter can cause the equipment to overheat or even malfunction. If possible, use "sine wave output" inverters to run electronic equipment.

Regulation is also important. The voltage from generators and inverters varies a lot more than voltage from the ac utility grid. Generators made for running tools or motors are usually less expensive, but are poorly regulated. For radio use, check the specifications to be sure the output voltage is acceptable both at no load and full load.

Batteries

Batteries supply dc power in place of power supplies for portable radios, emergency power, and other uses where ac power isn't available or practical. Batteries use a chemical reaction to generate current. The battery is constructed so that the chemical reaction can't occur until there is a path or circuit for electrons to flow between the battery terminals. When the chemicals are "used up" and the reaction stops, so does the current. The types of chemicals also determine the voltage of the battery.

There are many different types of batteries, but they fall into three basic groups:

- *Disposables*—battery chemicals can only react once, then the battery must be discarded
- *Rechargeables*—chemical reaction can be reversed, recharging the battery

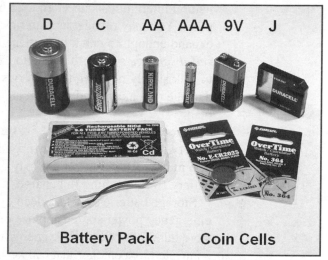

Figure 3-26—The photo shows several common sizes of batteries. Coin cells are usually used in radios as a source of backup power for the microprocessor circuitry. Battery packs are packages of several cells in a single enclosure or case. The photo shows a battery pack used for remote control vehicles. Courtesy Wiley Publishing, Two-Way Radios and Scanners for Dummies®.

Table 3-4
Battery Types and Characteristics

Battery Style	Chemistry Type	Full-Charge Voltage	Energy Rating (average)
AAA	Alkaline—Disposable	1.5 V	1100 mA-hr
AA	Alkaline—Disposable	1.5 V	2600-3200 mA-hr
AA	Carbon-Zinc—Disposable	1.5 V	600 mA-hr
AA	Nickel-Cadmium (NiCd)—Rechargeable	1.2 V	700 mA-hr
AA	Nickel-Metal Hydride (NiMH)—Rechargeable	1.2 V	1500—2200 mA-hr
AA	Lithium, Lithium Ion—Disposable	3.3—3.6 V	2100—2400 mA-hr
C	Alkaline—Disposable	1.5 V	7500 mA-hr
D	Alkaline—Disposable	1.5 V	14000 mA-hr
9V	Alkaline—Disposable	1.5 V	580 mA-hr
9V	Nickel-Cadmium (NiCd)—Rechargeable	9 V	110 mA-hr
9V	Nickel-Metal Hydride—Rechargeable	9V	150 mA-hr
Coin Cells	Lithium—Disposable	3—3.3 V	25—1000 mA-hr

• *Storage Batteries*—rechargeable batteries used for long-term energy storage

The most common types and sizes of disposable and rechargeable batteries used by hams are listed in **Table 3-4**. The column labeled "Chemistry" describes the chemicals used in the battery. The battery's *energy rating* in Ampere-hours (Ah) measures its ability to deliver current while still maintaining a steady output voltage. **Figure 3-26** shows several common types of batteries and their relative sizes.

To get the most energy from a battery, limit the amount of current drawn from it. A low discharge rate keeps the battery cool inside and minimizes losses due to the battery's natural internal resistance. To maximize battery life and capacity, store them in a cool, dry place. You may refrigerate batteries, but never freeze them since the resulting ice may expand enough to crack the battery. Heat accelerates the battery chemical's tendency to *self-discharge* so that they can no longer deliver as much charge. Moisture allows charge to leak slowly between the battery's external terminals. Regularly inspect batteries for damage or leakage and perform an occasional maintenance charge as part of your battery plan.

When recharging batteries, be sure to use a *charger* designed for that particular type of battery. Each different battery chemistry and size requires a certain method of charging. Too much recharging current may damage the battery. Too little current may keep the battery from reaching full charge.

Storage batteries, such as deep-cycle marine or RV batteries, are often used as an emergency power source in place of a power supply operating from ac power. They can store hundreds of times the energy in a small battery. Storage batteries are often left connected to a charger that can keep them fully charged with a small current, called *trickle* or *float charging*. Be sure that the charger will automatically switch to this lower current or it can overcharge and ruin these expensive batteries.

Storage batteries hold a lot of energy and must be treated with respect. They contain strong acids that can be hazardous if spilled or allowed to leak. Recharging storage batteries can also release or *vent* hydrogen gas, which is flammable. Be sure to charge them in a well-ventilated place. Accidentally short-circuiting a storage battery with a tool or faulty wiring can easily cause a fire and damage the battery.

Some batteries are actually made up of several individual packages of chemicals, called *cells*, connected in series, adding the voltages from each cell. The common "9 Volt" battery is actually six smaller 1.5-volt cells connected in series. An automobile's 12-volt battery is made of six compartments, each holding the necessary chemicals to form a cell that produces 2 volts. *Battery packs* are packages of several individual rechargeable batteries connected together in a single package and treated as a single battery.

When done with a battery, don't just throw it in the garbage. The materials in batteries are mildly toxic at best, so check at hardware stores or with your local government to see if there isn't a battery recycling program available.

Before you go on, study test questions T4C06-10; T5A05; T0A10, 11. Review this section if you have difficulty.

3.4 Handheld Transceivers

Handheld transceivers (handhelds) are incredibly popular and offer a variety of useful features. Because they can be carried by the operator and used while in motion, they're particularly useful for emergency communications. Even if you purchase a mobile rig, having a handheld radio for use away from the car is a good idea.

Dual-band handhelds can operate on two frequency bands—usually 2-meters and 70 cm. *Multi-band* handhelds can operate on several bands, adding coverage 6 or 1.25 meters or even 33 cm. Many handhelds also feature the ability to receive frequencies for other services, such as public safety, aircraft, or broadcast stations. This is called *extended* or *wide-band receive*. Check the radio's specifications for the exact frequency coverage.

Remember that it's not legal to use radios sold for Amateur Radio on other frequencies. Your radio may have the ability to listen outside the ham bands, but aside from the special Military Affiliate Radio System (MARS) frequencies, you can't transmit except in an emergency.

Using the handheld's built-in microphone and speaker requires the operator to hold the radio close to his or her mouth and ear. This may not be a problem, but occasionally it is inconvenient. One solution is to use a *speaker-mike* and use the belt clip supplied with the radio to hold the radio on a belt or in your pocket. The speaker-mike can then be clipped onto a shirt or jacket where you can hear the radio and use the microphone without holding the whole radio, too. *Headsets* are also available for many rigs and provide completely hands-free operation. The headset will have to provide its own VOX circuit to activate the radio's PTT circuit, unless the radio has VOX built-in. (See 3.1 for a discussion of VOX.)

BATTERIES

Handhelds use battery packs—groups of individual cells connected together in series to form a higher voltage battery, resulting in more power output from the radio. Most battery packs use rechargeable cells and the radio will come with a simple charger to keep the pack fully charged. A more sophisticated *fast charger* or *drop-in charger* may be available that can charge the pack quicker and usually acts as a convenient stand for the radio.

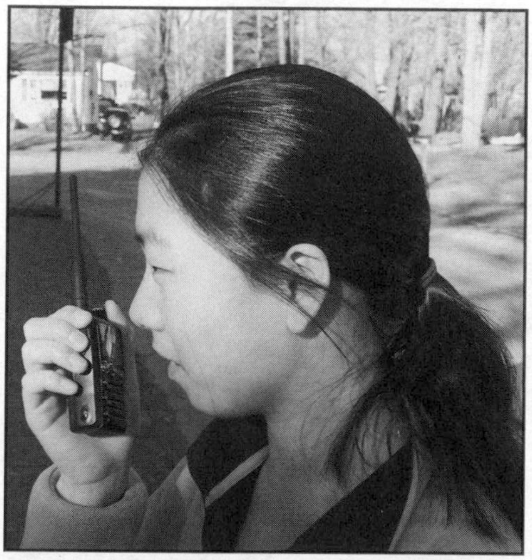

Rechargeable batteries or battery packs are convenient and less expensive than disposable batteries over time, but require a charger that operates from ac power. For operation in emergencies, disposable batteries are preferred because they do not depend on a battery charger for power. Many radios offer a battery pack for disposable AAA or AA batteries. If your interests include emergency communications or public service, be sure the radio select offers this option. In any case, you should have at least one spare battery pack.

Figure 3-27—To get the best range from a handheld radio, hold it so that the antenna is vertical. This aligns your antenna with those of a repeater or another handheld user. Also, turn the microphone slightly away from your face when talking so that your breath does not blow directly into it.

ANTENNAS

The flexible antenna used with most handheld radios is called a *rubber duck*. It's a ground-plane antenna that is shortened by coiling the conductor. The body of the radio and the operator form the antenna's ground plane. The rubber duck trades its convenient size against reduced performance compared to a full-sized ground-plane antenna.

A handheld radio can also be connected to a full-size or external antenna for better performance. The rubber duck can be detached and usually has a standard RF connector. This allows a mobile antenna to be used in a vehicle or the handheld to be connected to base station-style antennas at home.

While using the rubber duck antenna, for best performance hold the transceiver so that the antenna is vertical as shown in **Figure 3-27**. This aligns the handheld antenna with those of the repeater and most other handhelds. When antennas are misaligned, the received signals can be dramatically reduced, as much as 100 times! This happens because the polarization of the radio waves don't match that of the antenna. With mismatched polarizations, the effect of the radio wave on the receiving antenna is reduced.

Before you go on, study test questions T7A03; T9B07, 08, T9A04.
Review this section if you have difficulty.

3.5 RF Interference (RFI)

As more and more electronic devices and electrical appliances are put in use every day, interference between them and radios, called *Radio Frequency Interference* (RFI), becomes more commonplace. RFI can occur in either "direction," from or to the amateur radio equipment. Interference is made stronger with higher power or closer spacing to the noise source. In this chapter, we'll cover the basics of RFI so that you know what to expect when you encounter it and how to react.

FILTERS

Filters are an important part of radio, nowhere as important as in preventing and remedying RFI. Filters are used both to prevent unwanted signals from being radiated in the first place and to keep unwanted signals from being received. Filters are not only restricted to antenna feedlines, they can be applied to any conductor. To select the correct type filter, you have to know the nature of the interfering signals.

A filter that passes signals above a specified *cutoff* frequency is a *high-pass filter* (HPF) and below the cutoff frequency a *low-pass filter* (LPF). They *reject* or attenuate signals below and above their cutoff frequencies. *Band-pass filters* (BPF) reject signals outside the frequency band between the upper and lower cutoff frequencies. They act like a low-pass filter above the upper cutoff frequency and a high-pass filter below the lower cutoff frequency. **Figure 3-28** describes the action of each type of filter.

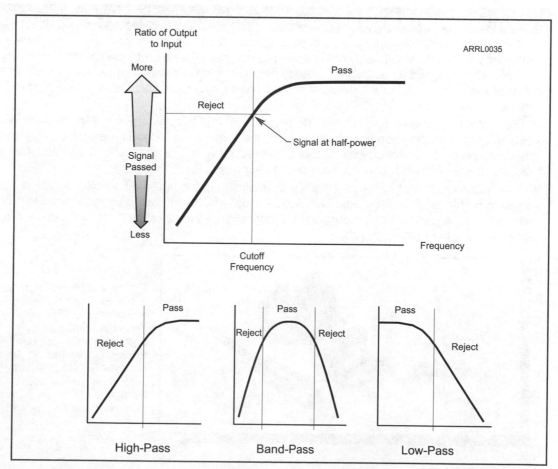

Figure 3-28—A filter's response graph shows how it affects signals at different frequencies. The vertical axis shows the ratio of the signal at the filter's output to the input. The horizontal axis shows frequency. The point at which the input signal is attenuated to half its original power is the cutoff frequency.

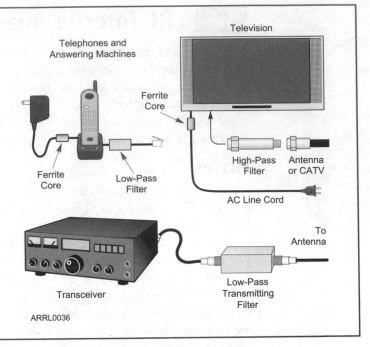

Figure 3-29—Filters are used in a number of different ways to reduce interference. Telephone or answering machine interference is often cured with a ferrite core placed on the power supply cable and a low-pass filter on the phone line. Both prevent RF signals from getting to the phone's electronics. A high-pass filter in the antenna or cable TV feedline often prevents strong amateur signals from enter the TV receiver and causing overload. Ferrite cores on the power cable keep RF from entering the TV by that route. A low-pass filter on the output of a transceiver prevents weak harmonics from being picked up by a radio or TV as interference.

Telephones and Answering Machines

Television

Ferrite Core

Ferrite Core

Low-Pass Filter

High-Pass Filter

Antenna or CATV

AC Line Cord

To Antenna

Transceiver

Low-Pass Transmitting Filter

ARRL0036

Ferrite—the RFI Buster

One of the most useful materials in dealing with RFI is the *ferrite core*. Ferrite is a ceramic magnetic material—you may have used ferrite magnets. The type of ferrite used for RFI suppression is specially designed to absorb RF energy over a broad frequency range, such as HF or VHF. Ferrite is available in many different *mixes* of slightly different composition that absorbs best in a particular range.

Ferrite is available as round *cores* (toroids), rods, and beads shown in **Figure 3-30**. Wires or cables are then wound on or passed through the ferrite forms. Beads are made large enough that they can be slipped over coaxial cables and secured with tape or a locking wire-tie. A wire or cable wound on such a ferrite core forms a choke filter.

One popular form of ferrite is the snap-core shown in the figure. The actual ferrite is a square block sawn in half through a large hole in it. A plastic case holds the two pieces together with a snap. This allows cords or cables to be wound on the core even if they already have connectors attached, such as power cords or video cables.

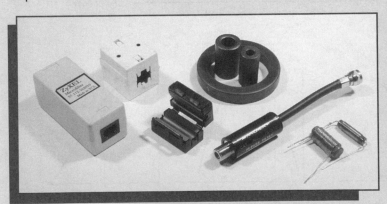

Figure 3-30—Ferrite is a ceramic magnetic material used in radio to make filters. It is available in many different forms; rings (toroids), rods, and beads. Cables can be passed through or wound on these cores to prevent RF signals from flowing along their outside surfaces. Courtesy Wiley Publishing, Two-Way Radios and Scanners for Dummies®.

Transmitting filters can pass high-power signals from a transmitter. Receiving filters are only intended for use in low-power receiver circuits. Low-pass filters that keep RF signals from passing into or out of equipment via the ac power connection are *ac power line filters*. They reject all signals with frequencies greater than a few kHz. *Choke* or *common-mode* filters are used to prevent RF currents from flowing on the outside of shielded cables, speaker wires, and other exposed conductors. **Figure 3-29** shows how filters are applied for several common types of RFI.

DIRECT DETECTION

A device need not be a receiver to experience interference. Some devices, such as telephones, music players, touch-sensitive devices, and other electronics can be affected by a strong RF signal. If RF signals can gain entry to an electronic device, components such as transistors, diodes, and ICs turn them into voltages and currents that affect the device's function, possibly upsetting its operation or distorting an audio signal. This is called *direct detection*. This is the most common form of interference to telephones, since they are rarely designed to include interference protection components.

To eliminate RFI caused by direct detection, the RF signal must be prevented from entering the equipment. Power cords, speaker or headphone cables, and accessory cables can all act as antennas to carry the RF signal to the internal circuits. RF filters and ferrite chokes can be added to cables to prevent RF pickup and detection.

The symptoms of direct detection are thumps or pulses when a transmitter is turned on and off. A garbled voice might be heard during AM or SSB transmissions. Strong FM signals are usually detected as hum.

OVERLOAD

Also called *fundamental overload*, overload is the result of strong signals overwhelming a receiver's ability to reject them. This creates interference to a desired signal inside the receiver. To eliminate overload, a filter or attenuator must be used at the input of the receiver to reduce the strength of the undesired signal.

A high-pass filter can be connected at the input of FM and TV receivers as shown in Figure 3-29 to reject strong lower-frequency signals from amateur HF and Citizen's Band stations. A low-pass filter at the amateur's transmitter will not solve overload problems—the problem lies within the receiver. As such, it is the receiver owner's responsibility to solve overload problems from a properly functioning transmitter.

Both consumer and amateur receivers can experience overload from nearby broadcast stations. *Broadcast-reject* filters attenuate signals from nearby AM, FM, or TV broadcast stations. The type of signal to be rejected must be specified when purchasing the filter since those broadcasts are on very different frequencies.

Symptoms of RFI from overload are interference across an entire band or many channels. If adding attenuation (either by turning on a receiver's attenuator or removing an antenna) causes the interference to disappear, it's likely due to overload.

HARMONICS

Due to minor imperfections, every transmitter's RF output signal contains weak harmonics of the desired output signal that can cause interference to nearby equipment. In extreme cases, a misadjusted or defective transmitter can generate strong harmonics. To eliminate harmonics a low-pass or band-pass filter is used at the output of the transmitter at the connection to the feedline as shown in Figure 3-29.

As a matter of good practice amateur HF stations often use a low-pass filter to keep any VHF harmonics generated by the transmitter from reaching the antenna. Even if the transmitter is completely within FCC rules, a nearby TV receiver could still pick up the VHF signal and experience interference. Harmonics that cause interference cannot be

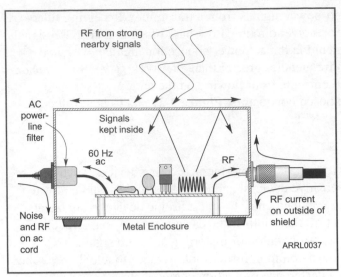

Figure 3-31—Shielding consists of a conductive surface surrounding an electronic circuit. Because RF signals cannot pass through the conductive material, only along it, signals can be prevented from leaking into or out of radio equipment.

filtered out at the receiver because they are on the same frequency as the desired signal. This is called *in-band interference*. Remember that filters to be used in a feedline carrying signals from a transmitter must be rated to carry the full transmitter output power.

SHIELDING

Even if all the wires and cables connected to electronic equipment were shielded, it would all be for naught if the circuits themselves could still receive and radiate signals. The circuitry must be protected from incoming RF and prevented from radiating RF signals of its own. This is why most sensitive electronics are enclosed in a metal enclosure—RF energy cannot pass through conducting surfaces. Surrounding a circuit with conducting material such as sheet metal or conductive paint or tape is called *shielding*. Shielding prevents RF energy from interfering with enclosed circuits. Shielding also prevents RF energy from escaping an enclosed circuit and possibly causing interference. **Figure 3-31** illustrates how shielding works.

Shielding not only protects enclosures, but also signals in cables, such as coaxial cable. Cable TV (CATV) systems, for example, can experience or cause interference when the shielding of the coaxial feed lines are broken or compromised. CATV channels are not the same as broadcast TV channels. Some CATV frequencies (inside the cable) are the same as amateur band frequencies (outside the cable). You can see that if the cable shielding isn't working, hams can interfere with CATV and vice versa. If a neighbor using cable TV suddenly begins experiencing RFI from your transmissions, it may be an indication that a feed line or connector has become loose or damaged.

NOISE SOURCES

Interference to amateur stations is not usually caused by a transmitter. It's far more common for amateurs to receive interfering noises from *unintentional radiators*. These signals are either leaked from electronic circuitry or generated as a by-product from electrical equipment. The following is a list of the most common "offenders."

- Electrical arcs from motors, thermostats, electric fences, neon signs, etc. generate raspy or clicking noises over a wide range of frequencies. The noise is strongest on the lower HF bands and gets weaker as frequency increases. The on-and-off pattern or rhythm of the noise is often a clue to what causes it. For example, electric fences generate a "pop" about once a second. AC power line filters on the offending equipment sometimes work.
- Power line noise from defective HV hardware has a characteristic 120 Hz buzz. Cracked or dirty insulators, loose connections, wires rubbing together—all these can cause power line noise. If you can track down the noise to a particular power pole or piece of equipment by using a portable broadcast or vehicle AM receiver, record the pole's identification numbers and call the power company. Do NOT attempt to shake or bump the pole—loose hardware can fail and drop a live wire right on you!
- Ignition noise caused by motor vehicles is a whine or buzz that varies with engine speed. It often only lasts while the vehicle travels past your location.

- Alternator whine is a type of noise caused by noise on the dc power system inside your own vehicle. You might hear it added to received audio but more likely, it will be heard by others on your transmitted signal. It is corrected with a dc power filter at your radio.
- Switching power supplies used by computers and consumer electronics generate mostly HF noise heard as unsteady tones at evenly spaced frequencies. Once the source has been identified, an ac power line filter is often effective.
- Computer and networking electronics may also directly generate signals at a single frequency, either steady or that vary in patterns. These can be very difficult to eliminate if the equipment is not shielded properly. Choke filters on all input and output cables are a good technique to start.

GUIDELINES

Dealing with interference is just a fact of life for hams. By preparing your own station and learning basic filtering techniques, you can keep interference from being too much of a problem. Remember that the FCC is a last resort for everyone. The FCC will require that everyone take all reasonable steps to identify and mitigate the effects of the interference before considering getting involved.

- Start by making sure your station is in good working order with appropriate grounding, filtering, and good quality connections, particularly for the RF signals.
- Eliminate interference to your own home appliances first. Demonstrating that you aren't interfering with your own devices is a good start. Eliminating interference at home is excellent practice, too!
- Eliminate sources of interference in your own home, such as worn out motor brushes, poorly filtered power supplies, and so forth. No only will it make operating more pleasant, it will be much easier to determine whether noise is caused elsewhere.

Speaking of elsewhere, you may eventually encounter a situation where your signals are causing interference to a neighbor or a device the neighbor owns is causing interference to you. Diplomacy is often required, even though your transmissions may not be at fault. Techniques for dealing with RFI to or from others is discussed on the ARRL's RFI Resources Web site at **www.arrl.org/tis/info/rfigen.html**, but remember these simple suggestions:

- Start by making sure it's really your transmissions that are causing the problem. It's not unknown for the mere presence of an antenna to generate a report of interference, deserved or not!
- Offer to help determine the nature of interference; detection, overload, harmonics. Knowing the cause leads to solutions.
- If you've determined that the noise is caused by a neighbor's equipment, offer to help determine the source of interference. Severe noise often indicates defective equipment that could be a safety hazard. Again, determining the source leads to solutions.
- Consult the ARRL RFI Resources Web site and printed material.

PART 15 RULES

Part 15 of the FCC's rules governs the responsibilities of owners of unlicensed devices that use RF communications (such as cordless phones or wireless data transceivers) or unintentional radiators, devices that radiate RF energy unintentionally as part of normal operation such as power lines, electric fences, or computers. The former are called *Part 15 devices*. In cases of interference to or from Part 15, you can begin to see the advantages of being federally licensed.

Reducing Part 15 to its basic principles:

- An unlicensed device permitted under Part 15 or an unintentional radiator may not cause interference to a licensed communications station, such as to an Amateur Radio station. Its owner must prevent it from causing such interference or stop operating it.
- An unlicensed device permitted under Part 15 must accept interference caused by a properly operating licensed communications station, such as from an Amateur Radio station.

What this means is that as long as your station is operating properly under the FCC's rules, then your operation is protected against interference by and complaints of interference to unlicensed equipment. If your signals are interfering with a television or telephone, it is the owner's responsibility to eliminate it, even though you may assist them. Similarly, it is the owner's responsibility to eliminate interference caused by their device, even with assistance from you. These rules are printed in the owner's manual for all unlicensed devices and are available on the FCC Web site (**www.fcc.gov**).

Before you go on, study test questions T5A06, 07; T3D02, 03, 07, 11; T5D01-10; T9B03. Review this section if you have difficulty.

Chapter 4

Communicating with other Hams

In this chapter, you'll learn about:

- **Common contact elements**
- **What contacts consist of**
- **Where contacts are made**
- **How to start and conduct a contact**
- **How to make contacts on a repeater**
- **What nets do**
- **How to find nets**
- **Emergency operating rules**
- **Amateur emergency organizations**
- **Special operating techniques and modes**

Having learned all that interesting material about rigs and electronics and radio waves, you know a lot about the technology of radio. In this section we turn to operating—how are contacts made and what does a contact consist of? We begin with the elements common to nearly every casual ham radio contact. Once you know these, you'll learn how the ham bands are organized so that you know where to tune the radio. You'll then discover how ham radio is conducted using repeaters and in the organized activities called "nets", especially during emergencies. This chapter concludes with coverage of a few of ham radio's many specialty activities. Clear your throat and get ready for that first contact!

4.1 Contact Basics

This part of the book is about "good amateur practices." There aren't many FCC rules about operating procedures, so amateurs have developed their own. They work pretty well, so it's a good idea to learn and practice them. The FCC (and other amateurs) expects you to follow established practices fairly closely, just as you expect other people to do in daily life.

CONTACT CONTENTS

The simplest possible ham-to-ham contact is an exchange of call signs—you can't get much simpler than that! Most contacts have a bit more to them—from casual conversations to coordinating a public event to formal radio messages. Yet all have some common elements. Some of the conventions and procedures radio operators use may be unexpected, but they have been developed as necessary for two or more people to communicate by radio.

General Principles

Remember that having a radio contact is different from a face-to-face meeting. When talking by radio, you can't see the other person, so all of the hand and facial movements we use to help us communicate aren't present. The only thing you can use is the audio that comes out of the radio (or characters that are displayed on the computer screen). Only one of you can talk at a time. To succeed at communicating, you'll need to learn "radio manners."

The most important item is to *identify* regularly on the air. Remember that except for your voice, you're invisible. Your radio name is your call sign, so use it whenever you announce your presence or when participating in a conversation. It's not enough to just identify yourself to the person you're in contact with—others listening need to know who you are, too! They may not know who "Bob", "Carol", or "James" are. Give your call at the beginning of and frequently throughout the contact. Then identify once more at the contact's end.

As you converse, you should also make it clear when you're done transmitting. Remember, the listeners can't see you, so they can't tell if you're done talking or just thinking. That's why phone operators say, "Over" and a telegrapher sends "K" or "$\overline{BK}$"—so the other party knows it's their turn. When the contact is completely done, "Clear" or "$\overline{SK}$" is used to let others know that the frequency is available for a contact. These abbreviations are *procedural signals* and have been part of contacts even back in the days of the telegraph before radio!

Take pains to speak clearly and a little slower than in normal conversation. Remember, all you can hear comes through the radio speaker—no eye contact! That means if you mumble or rush or slur during voice contacts, you will be hard to understand. The same goes for Morse code—send carefully and take pride in forming your words clearly by sending at a comfortable speed. A CW operator whose code is crisp and easy to read is said to have a good *fist*—their on-the-air signature. Even on the digital modes, efforts to spell and punctuate correctly are appreciated.

You'll also find that hams use a lot of jargon such as abbreviations, procedural signals, acronyms, and slang. These can be hard to figure out on your own but are important to learn because they save time, communicate effectively, and help you fit in. Your Elmer will help you understand them and it's also a good idea to have a reference book handy, such as those listed in the first chapter. With a little practice you'll soon sound like an OT (Old Timer).

Using a Frequency

After you've been on the air for a while, you'll notice that hams tend to use the same frequencies on a regular basis. For example, NØAX might "hang out" on the 443.50 repeater or use CW on 7.035 MHz. No matter how much NØAX stays on that frequency, though, no station has exclusive rights to any particular frequency. Even groups that meet on the same frequency every day at the same time have no priority right to use it. If another ham is using the frequency, they should be prepared to move to another frequency or wait. While it's easy to see how this could lead to difficulties, most hams realize the need to be flexible and are accommodating of the needs of other hams. The key is to be polite and always have a Plan B for making your contacts.

Signal Reports

Immediately after establishing contact, it is customary to let the other station know how well their signal is being received. This is important because the other station can then compensate for poor conditions by sending or speaking

Table 4-1
ITU Phonetic Alphabet

Letter	Word	Pronunciation
A	Alfa	**AL** FAH
B	Bravo	**BRAH** VOH
C	Charlie	**CHAR** LEE
D	Delta	**DELL** TAH
E	Echo	**ECK** OH
F	Foxtrot	**FOKS** TROT
G	Golf	GOLF
H	Hotel	HOH **TELL**
I	India	**IN** DEE AH
J	Juliett	**JEW** LEE ETT
K	Kilo	**KEY** LOH
L	Lima	**LEE** MAH
M	Mike	MIKE
N	November	NO **VEM** BER
O	Oscar	**OSS** CAH
P	Papa	PAH **PAH**
Q	Quebec	KEH **BECK**
R	Romeo	**ROW** ME OH
S	Sierra	SEE **AIR** RAH
T	Tango	**TANG** GO
U	Uniform	**YOU** NEE FORM
V	Victor	**VIK** TAH
W	Whiskey	**WISS** KEY
X	X-Ray	**ECKS** RAY
Y	Yankee	**YANG** KEY
Z	Zulu	**ZOO** LOO

Note: The **boldfaced** syllables are emphasized. The pronunciations shown in this table were designed for those who speak any of the international languages. The pronunciations given for "Oscar" and "Victor" may seem awkward to English-speaking people in the US.

Table 4-2
The RST System

READABILITY

1—Unreadable.

2—Barely readable, occasional words distinguishable.

3—Readable with considerable difficulty.

4—Readable with practically no difficulty.

5—Perfectly readable.

SIGNAL STRENGTH

1—Faint signals barely perceptible.

2—Very weak signals.

3—Weak signals.

4—Fair signals.

5—Fairly good signals.

6—Good signals.

7—Moderately strong signals.

8—Strong signals.

9—Extremely strong signals.

TONE (CW and Digital)

1—Sixty-cycle ac or less, very rough and broad.

2—Very rough ac, very harsh and broad.

3—Rough ac tone, rectified but not filtered.

4—Rough note, some trace of filtering.

5—Filtered rectified ac but strongly ripple-modulated.

6—Filtered tone, definite trace of ripple modulation.

7—Near pure tone, trace of ripple modulation.

8—Near perfect tone, slight trace of modulation.

9—Perfect tone, no trace of ripple or modulation of any kind.

The "tone" report refers only to the purity of the signal. It has no connection with its stability or freedom from clicks or chirps. Most of the signals you hear will be a T-9. Other tone reports occur mainly if the power supply filter capacitors are not doing a thorough job. If so, some trace of ac ripple finds its way onto the transmitted signal. If it has a chirp or "tail" (either on "make" or "break") add C (for example, 469C). If it has clicks or noticeable other keying transients, add K (for example, 469K). Of course a signal could have both chirps and clicks, in which case both C and K could be used (for example, RST 469CK).

more slowly or repeating items. Conversely, if signals are strong, unnecessary repetitions can be avoided. This information is called a *signal report* and is exchanged in one of two formats.

"RST" is used during SSB, digital and CW contacts. The letters stand for Readability, Strength, and Tone and the report is sent as a series of three numbers, such as 339, 599, or 457. Readability has a value of 1 to 5 and rates how well the signal can be understood. Strength is measured on a scale of 1 to 9 and is often reported as the receiver's peak S-meter reading. (Readings greater than S9 are simply reported as 9 or "S9" plus whatever the meter reading shows.) Tone is only used for CW transmissions and refers to the quality of the signal. Clear, pure tones that have a steady frequency or with no audible hum rate a 9—as do most of the signals produced by modern radios. **Table 4-2** shows all the variations of this popular reporting system. Voice contacts also use the "Q" or "Quality" system. A number from 1 (meaning barely understandable) to 5 (meaning perfectly readable) follows the Q.

Power Level

It's also nice to know how much power the other station is "running" along with the signal report. If a QRP (low power) station's signal is worthy of an RS of 59, that's an accomplishment! Power is not generally of that much interest during repeater contacts, since the signals are relayed.

The FCC rules are quite clear that hams should use the minimum amount of power needed to make the contact. That doesn't mean you should reduce power until your signal can just barely be heard. The intent of the rule is to use an appropriate power level that does not deny others the use of the frequency farther away—more a concern on SSB, digital and CW than it is on repeaters, where the repeater output power can't be adjusted. In general, if your signal is perfectly understandable while running "barefoot" (meaning without an external power amplifier), don't turn the amplifier on. As always, be sure to comply with the legal power limits in the rules!

Locators

You have exchanged signal reports and power, so it is now time to find out where the other station is. The exact way to identify location varies, from the local mobile station's "I'm on the Beltway, just south of the toll road" to "Location is latitude 43 North and longitude 39 West" of a radio officer on a freighter in the Atlantic. Location can always,

Table 4-3
Q Signals

These Q signals are the ones used most often on the air. (Q abbreviations take the form of questions only when they are sent followed by a question mark.)

QRG   Your exact frequency (or that of ___) is ____kHz.
      Will you tell me my exact frequency (or that of _____)?

QRL   I am busy (or I am busy with ____). Are you busy? Usually used to see if a frequency is busy.

QRM   Your transmission is being interfered with ____
      (1. Nil; 2. Slightly; 3. Moderately; 4. Severely; 5. Extremely.)
      Is my transmission being interfered with?

QRN   I am troubled by static ____. (1 to 5 as under QRM.)
      Are you troubled by static?

QRO   Increase power. Shall I increase power?

QRP   Decrease power. Shall I decrease power?

QRQ   Send faster (____wpm). Shall I send faster?

QRS   Send more slowly (____wpm). Shall I send more slowly?

QRT   Stop sending. Shall I stop sending?

QRU   I have nothing for you. Have you anything for me?

QRV   I am ready. Are you ready?

QRX   I will call you again at ___hours (on ___kHz).
      When will you call me again? Minutes are usually implied rather than hours.

QRZ   You are being called by ____ (on ___kHz).
      Who is calling me?

QSB   Your signals are fading. Are my signals fading?

QSK   I can hear you between signals; break in on my transmission.
      Can you hear me between your signals and if so can I break in on your transmission?

QSL   I am acknowledging receipt.
      Can you acknowledge receipt (of a message or transmission)?

QSO   I can communicate with ____ direct (or relay through ___).
      Can you communicate with ___ direct or by relay?

QSP   I will relay to ___. Will you relay to ___?

QST   General call preceding a message addressed to all amateurs and ARRL members.
      This is in effect "CQ ARRL."

QSX   I am listening to ___ on ___kHz. Will you listen to ___on ___kHz?

QSY   Change to transmission on another frequency (or on ___kHz).
      Shall I change to transmission on another frequency (or on ___kHz)?

QTC   I have ___messages for you (or for ___). How many messages have you to send?

QTH   My location is ____. What is your location?

QTR   The time is ____. What is the correct time?

Q-Signals

Q-signals are a system of radio shorthand as old as wireless and developed from even older telegraphy codes. Q-signals are a set of abbreviations for common information that both save time and also allowed communication between operators that didn't speak a common language. Modern ham radio uses them extensively. **Table 4-3** lists the most common Q-signals used by hams. While Q-signals were developed for use by Morse operators, their use is common on phone, as well. You will often hear, "QRZed?" as someone asks "Who is calling me?" or "I'm getting a little QRM" from an operator receiving some interference or "Let's QSY to 146.55" as two operators change from the calling frequency to a nearby simplex communications frequency.

however, be denoted or queried with the *Q-signal* QTH. "My QTH is…" means, "My location is…" and "QTH?" means, "What is your location?" (See the sidebar on Q-signals for more on these abbreviations.)

An increasingly popular method of identifying location is the Maidenhead Locator System, better known as grid squares. In this system, named for the town outside London, England where the method was first created, the Earth's surface is divided into a system of rectangles based on latitude and longitude. Each rectangle is identified with a combination of letters and numbers. A four-digit code of two letters followed by two numbers identifies a unique rectangle of 1° latitude by 2° longitude. For example, ARRL Headquarters in Newington, Connecticut is located in grid square FN31. A further two letters can be added for greater precision, such as FN31pq for the precise location of the ARRL station. (Learn more about grid squares at **www.arrl.org/locate**.)

Appropriate Topics

As you listen to amateurs discussing everything from the weather to sporting events and public service, you'll wonder what hams don't talk about? Hams pride themselves on professional, high-quality procedures and conduct, but there are really very few restrictions about what you can talk about over the air.

No matter what the topic, indecent and obscene language is flatly prohibited. No cussin' allowed! Just don't! If you find yourself getting upset for any reason, it's best to just turn off the radio for a while. While there is no official list of "the dirty words you can't say on the radio," use good sense and an extra helping of manners to avoid offending other hams. That especially includes racial and ethnic references. Everything you say travels a long way and you never know who might be listening!

Similarly to the use of language, hams also try to stay clear of provocative subjects that often create strong feelings; politics, religion, and sexual topics. There are plenty of forums for these discussions, such as the Internet and e-mail. Ham radio is for developing communications expertise and goodwill. If the discussion doesn't further those aims, it's probably best taken elsewhere.

Signing Off

When your contact is complete concluding it is simple, but like in real life, it's rude to just "hang up." Hams have their own way of saying goodbye that can get to be somewhat involved. The best way to learn (like many other things in ham radio) is to listen. Here are a few of the terms you'll hear as hams go their separate ways:

- Final—the last transmission, as in "I will be clear on your final."
- QRU—the Q-signal that means, "I have nothing more for you"
- Down the log—"I'll see you later"
- 73—"Best Regards," an almost universal closing motif
- 88—added for a female operator, originally "Love and Kisses"

As you complete your final transmission of a contact and do not intend to respond, it is customary to add "Clear" (or CL in Morse) to let everyone know you are through. Occasionally you may hear "Out" as in "Over and out", but that is mostly a relic of the movies. "Clear" or "Off and clear" is unambiguous.

ADVISING AND ASSISTING

The longest tradition of Amateur Radio is that of mutual assistance. A century ago there were no radio texts or handbooks and so all hams were self-taught, relying on others to help them get on the air and stay there. The same ethic is alive and well today. While as a new ham you may be in need of assistance today, you'll be glad to help another ham tomorrow.

Methods and Procedure

It's important to start by noting that every ham was once at the same level of learning that you are today. Crack operators and technical whizzes are not born, they're made. Don't be embarrassed because you don't know everything about ham radio or make a mistake! There are plenty of ways to learn, others will help (particularly if you ask questions), and you'll learn rapidly. Accept criticism in the helpful spirit of its offering and be sure to extend the same spirit to others.

The most common mistakes made by any ham are ones of technique—transmitting too soon or too late, using the wrong procedure, misunderstanding instructions, etc. While on the air, be sure to take a helpful tone in correcting the operator. Tell them what you expected and give guidance as to what to do next time. For example, if a station makes a call on a repeater in the middle of a public service exercise, perhaps they just didn't listen long enough before transmitting. You may be annoyed, but don't let that color your response. An appropriate response would be something like, "W1AW, this is K1ZZ net control for the Podunk Valley Parade. The repeater is being used to coordinate the parade until about 1 o'clock. You could use the 145.35 repeater, OK?" This avoids prejudging W1AW's error, let's them know what's happening and makes a helpful suggestion. This will generally correct the situation promptly and avoids hurt feelings.

With all the features of modern radios, it's also easy to get confused and transmit in a way you didn't expect, such as on the wrong mode or frequency. (Hint—that's why you might want a simple radio to start with instead of one with all the bells and whistles.) Even experienced hams goof now and then and you will, too! If you hear someone having trouble operating their radio, such as transmitting off frequency or using FM in the SSB part of the band. Use the same technique to correct them as in the preceding example. Contact them and let them know what the problem is. If you can, offer a solution; "Your voice is distorted, try speaking more softly." A helpful, friendly voice when you're having trouble is appreciated!

Radio and Antenna Checks

There will be times when you are not sure your radio is working properly. When you need an on-the-air evaluation of your signal, that's called a *radio check*. Before asking for help, make sure you can clearly describe the problem and what kind of evaluation you need. If you can arrange for a friend to meet you on the air that's the best method of troubleshooting.

If a friend's not available, you may need to make a general call. The technique is to find an ongoing contact and break in as above. When the stations acknowledge your call sign, you respond with, "This is W1AW and I need a radio check, please." If the stations can help you, they'll take it from there. Be sure to identify yourself with your call sign and be polite. If they are busy and can't help, move to another frequency. You can also make the same call on an unused frequency where someone may hear you and respond. If possible, don't use a busy repeater to do the tests.

If you are asked to give a radio check, listen carefully and respond with detailed information. Distortion of voice signals could be microphone or RF feedback problems. Excess noise or crackling indicates a broken or intermittent connection. Hum on any type of signal might indicate power supply or battery problems.

Antenna comparisons are the next most common testing need. Hams are always experimenting or replacing antennas in search of a better signal. The same procedure applies for radio checks; pay attention to details and don't hesitate to relay your observations to the testing station. This is the time for an accurate and precise signal report! If you are conducting the test, be sure to give the evaluating station enough time to observe the antenna's performance with a *five-count* where you count on the air, "One-two-three-four-five" or on CW send "V" several times or a few seconds of unmodulated signal. Always be sure to properly identify your station!

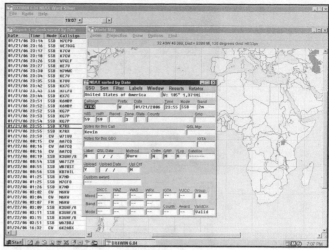

Figure 4-1—Software logging programs can store and cross-reference thousands of contacts. For use away from a computer, the paper logbooks are a reliable standby and the information can be transcribed into a computer later.

Noting Violations

Hams have a long-standing tradition of self-policing to help each other play by the rules. Part of that tradition is notifying each other of apparent infractions. You may also come across a station violating the Amateur Service rules, such as transmitting on voice in a CW-only band segment. Don't commit an infraction yourself! Contact the station by mail or phone and let them know. Be diplomatic and keep your tone helpful.

The ARRL has also established the Official Observer (OO) program (**www.arrl.org/ hrlm/section-4.html**) that relies on technically skilled hams. The OOs keep an ear on the bands, often finding problems before they become a problem for other hams.

LOGGING AND CONFIRMING CONTACTS

Hams are no longer required by the rules to keep a logbook, but it is an excellent idea to do so. There are a variety of paper logbooks and software logging programs available from most ham radio vendors. Enter *radio logging software* into an Internet search engine and you'll find plenty of choices. Why keep a log? It's useful to have a record of contacts with the hams you meet on the air. Keeping a log can also help you identify good times to operate or help identify sources of interference, both to and from your station. It's your record of how you make use of your license!

Most hams log random simplex contacts made on HF or VHF/UHF bands while at home or operating in a portable location, such as during a vacation or even a radio *DXpedition* to an unusual spot. Casual and regular contacts made with friends aren't usually logged, nor are most local repeater contacts. Contacts made during mobile operation, which requires extra attention to the hazards of driving, are rarely logged.

When keeping a log, record the time and date of the contact, the band and mode, and the call sign of the station you contacted. These are the minimum needed to confirm a contact later. You might want to record the exact frequency, your power level, signal reports exchanged, and information about the other station such as location and operator name. Some logging software can connect directly to your radio and record much of that information for you!

Figure 4-2—Exchanging QSLs after a contact is a cherished part of ham radio for many hams. The cards can be simple or ornate, text-only or beautiful photos. With the availability of excellent home color printers, many hams design and print their own cards.

Many hams exchange cards after a contact, particularly on the HF bands or when using SSB and CW. (Repeater contacts are rarely acknowledged in this way.) Nearly every ham has a personalized card like those in **Figure 4-2**, called a *QSL* after the Q-signal that means "received and understood." Sending the cards is called *QSLing*. Hams collect the cards just for personal interest or to submit as proof-of-contacts for award programs such as those listed at **www.arrl.org/hrlm/section-4.html**.

You can create your own card with a home computer or order them from several vendors. If you decide to make your own, get a few cards from friends and study them to be sure you include all the necessary information as described above. Make sure to include an accurate description of your location, such as county and grid square, since many award programs require this information. There are also electronic QSL systems such as the ARRL's Logbook Of the World. Most logging software can automatically transfer your contact information to these electronic systems, saving postage and time. Nevertheless, a paper QSL arriving by mail has a special meaning for many hams.

Before you go on, study test questions T3B07, T3C04-10, T6C10, 11, T7A09. Review this section if you have difficulty.

4.2 Band Plans

The amateur bands may not look like a lot of room on a big frequency chart, but once you start tuning around, you'll find they are quite large! How can you tell where to tune so that you can find your preferred activity? What keeps the different types of operating from being sprinkled randomly all over the band?

DEFINITIONS & FINDING BAND PLANS

The solution to both locating and organizing activity begins with a *band plan*. A band plan goes beyond the FCC rules and regulations that segregate the amateur bands by mode and license class. Band plans always comply with the FCC rules—they just provide some additional structure. Band plans are voluntary agreements by hams and don't have the same authority as a rule. However, band plans are considered "good amateur practice" and the FCC expects hams to follow their suggestions unless they have a good reason not to. There are practical reasons to follow the band plan, as well—you won't make many contacts using a handheld FM radio if you transmit on a frequency used by SSB operators!

The easiest place to find the band plans for U.S. amateurs is via the ARRL's home page or on the *Ham Radio License Manual* Web site at, **www.arrl.org/hrlm/section-4.html**. You should take a few minutes to browse the bands plans, particularly those for the 2-meter and 70-cm bands where many Technician-class licensees begin operating. **Table 4-4** shows the

Table 4-4

2-meter (144-148 MHz) Band Plan

144.00-144.05	EME (CW)
144.05-144.10	General CW and weak signals
144.10-144.20	EME and weak-signal SSB
144.200	National calling frequency
144.200-144.275	General SSB operation
144.275-144.300	Propagation beacons
144.30-144.50	OSCAR subband
144.50-144.60	Linear translator inputs
144.60-144.90	FM repeater inputs
144.90-145.10	Weak signal and FM simplex (145.01,03,05,07,09 are widely used for packet)
145.10-145.20	Linear translator outputs
145.20-145.50	FM repeater outputs
145.50-145.80	Miscellaneous and experimental modes
145.80-146.00	OSCAR subband
146.01-146.37	Repeater inputs
146.40-146.58	Simplex
146.52	National Simplex Calling Frequency
146.61-146.97	Repeater outputs
147.00-147.39	Repeater outputs
147.42-147.57	Simplex
147.60-147.99	Repeater inputs

2-meter band plan. You'll probably be surprised at all the different types of activity!

On the HF bands, band plans tend to be simpler because all activity is simplex (except for repeaters on the 10-meter band). On the VHF and UHF bands where repeater operation is common, it's necessary to show where repeaters listen as well as transmit. Areas of those bands are also set aside for simplex operation. Here are some definitions of the band plan terms:

- DX Window—because bands are often full of local or regional contacts, DX windows give the weaker signals traveling a long distance a relatively clear space to be heard. DX contacts can certainly be made elsewhere, but the windows are good places to start tuning.
- Digital modes, RTTY (Radioteletype), Packet—Digital operation is not as popular as voice and CW, so clustering the signals together makes it easier to find other digital mode users.
- Beacons—Automated transmissions are used to tell when the band is "open" to the area of the world in which the beacon is located. The Northern California DX Foundation (**www.ncdxf.org**) operates a series of beacons around the world on the HF bands and many individual hams operate a beacon on 10-meters or the VHF/UHF bands.
- Weak Signal—"Weak" is used as opposed to the customarily strong signals from FM repeaters. CW and SSB are used in the Weak Signal parts of the band because those modes allow communication at lower signal strengths.
- Satellite Uplinks and Downlinks—These are the segments of the bands where signals are sent to (*uplink*) and received from (*downlink*) satellites.
- Simplex, Repeater Inputs and Outputs—This is where the usual repeater operations are found and where nearly all FM operation occurs.
- Control links—Repeaters and other stations that are linked together or controlled by a remote operator use radio links to carry audio and control signals.

CALLING FREQUENCIES & BEACONS

Nearly every band has one or more *calling frequencies* listed. These are frequencies on which users of a specific mode or style of operation make contact. They then tune to a nearby frequency where they continue the contact. A calling frequency allows stations to find each other when continuous tuning over a larger range would be unproductive. These are also very useful for detecting unusual propagation by continuously monitoring the calling frequency and making the occasional CQ. A sudden change in propagation will then be noticed by many amateurs instead of one lucky operator who just happened to be on the right frequency.

WHY BAND PLANS ARE NEEDED

You are probably getting the idea behind a band plan. These voluntary conventions simply make it easier for everyone to maximize their own success and share the some-times-crowded ham bands. Sharing is particularly important because the different styles of operating are not always compatible. For example, a digital station and CW or voice station simply can't share a common frequency—they are too different. It's a lot easier for signals of the same type to operate shoulder-to-shoulder than a mix of modes.

The other major reason for having band plans at VHF and above is the effective use of repeater stations. As we mentioned in the previous section, repeaters have an input and an output frequency separated by the repeater offset frequency. Having the repeater input frequencies close together and the output frequencies close together serve two purposes. It's a lot easier to tune from repeater to repeater that way and by keeping the input frequencies separate from the powerful output signals, sensitivity is improved. Everything works a lot better when the band plan is followed!

WHO MAKES BAND PLANS

You might think that the ARRL has cooked up all these band plans, but that's not the case. The ARRL publishes what it understands to be the band plan based on how hams use the various frequencies. These voluntary "gentlemen's agreements" evolved over the years as the various groups of hams decided how best to coordinate their operating. For example, the international club of (QRP) low-power operators forged an agreement to congregate around a specific set of frequencies on all the bands and promoted them. The ARRL band plans simply reflect those agreements.

Repeater band plans at VHF and UHF were a bit more difficult to organize, since each region was isolated by limits to signal propagation. This often had a big impact on decisions about where to place input and output frequencies and the *channel spacing* between one repeater and the next. Fortunately, it was quickly realized that a certain amount of compatibility between the plans was needed. Hams have almost completely aligned the regional repeater band plans so that travelers can use their FM handheld radios anywhere.

BAND PLAN RULES

As mentioned earlier, band plans are considered "good practice" by the FCC and so it expects hams to abide by them. The only time this really comes into question is when persistent interference is caused by a station operating in conflict with the band plan. For example, if someone decides to operate a repeater whose output frequency is the same as a nearby repeater's input frequency, interference will result. If the parties are unable to resolve the issue with the help of the regional *frequency coordinator* committee, the FCC will first look to see which repeater is operating according to the band plan.

Band plans by themselves can not guarantee that no interference exists. For example, on 40-meters, the international RTTY calling frequency is the same as that of the QRP CW operators! The powerful RTTY and weak Morse code signals don't coexist very well. Luckily, it is relatively uncommon for interference between the two groups to last very long. Hams, being much more flexible or *frequency-agile* than any other radio service, just retune a bit and carry on.

Study test questions T3B01-03. Review this section if you have difficulty.

4.3 Making a contact

The very first thing to do and most important for the new ham, is to be sure your license gives you permission to use that frequency and mode before you transmit! As a Technician licensee, you will have full amateur privileges on all amateur bands of 50 MHz or higher, but there are restrictions within those bands as to mode.

Once you have determined that you're authorized to operate on a certain frequency and mode, you have three ways to make contacts. You can wait for someone to come along to your listening frequency, you can tune your transceiver looking for someone to contact, or you can make a general call that will announce you want to make a contact. The first is called *monitoring a frequency*. The second is called *tuning the band*. The third is to solicit a contact. On SSB or CW this is done by *calling CQ*. (CQ is another procedural signal that means "I am calling any station.") Repeater contacts are initiated slightly differently as described below.

> The very first thing and most important, be sure your license gives you permission to use that frequency and mode before you

Once you've successfully joined or initiated a contact, you may find yourself in a friendly extended conversation, called a *ragchew*. If you're sharing the frequency with several other stations, that's a *round-table*. Stations in a rare location may only exchange signal reports and call signs so that as many stations may make contact with them during their short visit. Organized on-the-air meetings are called *nets*. These are only a few of the activities that you'll find on the air.

STARTING A CONTACT

If you decide to solicit a contact, you should be sure that some other ham isn't already using the frequency. It's easy to do—just listen! Repeaters usually transmit a strong, clear signal, so it's pretty obvious if someone is using the repeater. If you don't hear someone in ten seconds or more, chances are good that the repeater is unoccupied. On SSB or CW, if you don't hear any signals in five to ten seconds an ongoing contact or activity may have paused or you may not be able to hear the station currently transmitting, so ask if the frequency is clear. Just say, "This is W1AW, is the frequency in use?" or on Morse code, "QRL? DE W1AW".

What if someone else is using the frequency or repeater and you accidentally interrupt them? Don't panic, just say "Sorry, W1AW clear" ("SRI W1AW CL" if using CW) and wait for their conversation or activity to conclude or tune to a different frequency.

On SSB or CW, a station calling CQ will actually send or say "CQ" several times followed by their call sign. It sounds like this:

W1AW: "CQ CQ CQ, this is W1AW Whiskey One Alpha Whiskey calling CQ and standing by."

If you hear a station calling CQ, it's easy to respond. Give their call sign once (they already know their own call!) and then yours once (if they are strong and clear) or twice. Give your call clearly and distinctly so that they can understand it if there is noise or interference. Your response should sound like this:

N6ZFO: "W1AW this is November Six Zulu Foxtrot Oscar, November Six Zulu Foxtrot Oscar, over."

Use *phonetics* (see Table 4-1) when using a voice mode, such as SSB or FM, so they get your call sign correct.

Because repeaters usually have a strong signal on a known frequency, it's not necessary to make a long general call (CQ) to attract listeners tuning by. The easiest way is just to announce that you are listening and are available for a contact. It sounds like this:

"W1AW is monitoring." Or "W1AW is monitoring and looking for a contact."

That announces to everyone listening that W1AW is available for contacts through the

"My name is Chris, Charlie Hotel Radio India Sierra." Why do hams (and radio operators in general) use those strange words instead of just spelling with letters? Remember that you only hear what comes through the radio, so it's often difficult to tell the letters apart. For example, "E", "T", "B", "C", "D" and so forth can sound a lot alike. Not only that, but non-English speakers don't pronounce letters in the same way at all! The solution to both of these problems was *phonetics*.

Developed in the early days of international radio, the International Telecommunications Union (ITU) developed a standardized *phonetic alphabet* that operators of all languages could use to exchange precise information. The alphabet is shown in **Table 4-1**. Each word was chosen because it was hard to confuse over the radio. Hams should learn these standard words and use them whenever there is a need for precise, exact spelling—for example, your call sign!

You may also be familiar with the US military version of phonetics and these are often encountered. Hams that specialize in round-the-world contacts sometimes use the names of countries or cities. For example, you might hear "Norway" instead of "November" or "Santiago" instead of "Sierra". There is no FCC rule about which set is required. Use the set that best suits the need.

Use caution in making up your own set of funny or cute words, such as "Wanted, One Aged Whiskey" for W1AW. Those may be fine in your local club since they're just nicknames, really, but they can be very confusing on the air, particularly to a foreign ham.

repeater. The exact method varies from region to region and even from repeater to repeater, so listen a while first to see how others make contacts and follow their lead. (Hint—listening to learn is a sure path to success in ham radio!") Respond the same as if the station had called CQ.

What if you are sure that your signal is being received by the repeater but the station you called didn't respond? Don't feel slighted! Remember that many stations use repeaters as a kind of on-the-air meeting place for club members or acquaintances. They may not be looking for a random contact at that time. For best success on repeaters, wait until an existing contact is finishing and then call one of the stations. This works especially well if you can discuss a topic of their just-concluded contact—a real conversation starter! Participating regularly in nets (discussed later) or other events on repeaters is a good way to break in with a group, too.

Operating on CW (Morse code) follows much the same rules, without the phonetics, of course! Remember to use the procedural signal "K" at the end of your transmission so that the station you're calling knows that you've finished. Here's a tip—if you're the one calling CQ, be sure to send at a speed you'd feel comfortable receiving! When responding to another's CQ, send at about the same speed.

Taking Turns and Breaking In

Once again recalling that radio communications can't rely on visual cues, there are certain methods that hams developed or adopted to control the flow of a conversation. The most common is the word "Over" or the procedural signal K for Morse. Hams learn not to begin speaking until the frequency has been released by the transmitting station. That's why you should say "Over" (or send K) when you have finished transmitting the current sentence. It lets the other station know that you are ready for a reply and are not just thinking or pausing. If you are having a conversation with good signal quality and are familiar with the other operators, the need to use "Over" every time is relaxed, but it is still very useful in keeping the conversation flowing.

Repeater receivers can easily sense when a station has stopped transmitting because the input signal disappears. As a useful service to its users, the repeater often adds a short *courtesy beep* to the re-transmitted signal when the transmitting station stops speaking. This becomes the "over" cue to other stations to start speaking, although "Over" is

The only exception to the "no one owns a frequency" rule is during an FCC Communications Emergency Declaration. During natural disasters or other communications emergencies, the FCC will make a declaration that certain frequencies or ranges of frequencies are reserved for emergency communications. On those frequencies, emergency communicators still have to work things out so they can share the available room. The reservation stays in effect until the FCC lifts the declaration. You can learn of these declarations from ARRL bulletins transmitted through W1AW or on the ARRL Web site at **www.arrl.org/w1aw**.

common on repeaters as well. Some radios have the capability to add their own courtesy beep, but it is not necessary (and often confusing) on a repeater with the feature already.

Even the best procedure can't guarantee that two stations won't accidentally start transmitting at the same time. Because ham transceivers can't receive while transmitting, there's no way to know of the simultaneous transmissions until one station stops. Two stations talking at once is called *doubling*. When that happens, someone listening in might say, "You doubled!" At that point, one station asks for a repeat transmission and waits.

If two stations are having a contact and you need to enter the conversation for some reason, it is necessary to get their attention to get an opportunity to speak. This is called *breaking in* and requires a little finesse in technique. The best way to break in is to wait for one station to stop transmitting and then quickly say, "Break [your call sign]." (BK is the equivalent on CW.) It is good etiquette during your contacts to pause briefly before pressing the microphone PTT switch to give a *breaking* station the chance to transmit. If you hear someone breaking in, the appropriate response is to pause and ask the breaking station to go ahead. For emergency break-ins, see the text below for a discussion of how to go about it. *Never* use an emergency signal when you don't need to!

REPEATER USE

As a Technician-class licensee, you are most likely to make contacts through a repeater. Because many amateurs listen to a repeater at the same time, repeater contacts have some operating procedures that are different than those used for SSB and CW contacts.

Finding Repeaters

First, how do you find a repeater that you can use? Start by looking at the band plan. Find the band segment filled with repeater output frequencies. You can tune through that part of the band, but there is no guarantee that you'll find all of the repeaters because they don't transmit all the time.

If you hear a repeater, how can you tell which one it is? Listen for a while and you may hear an automated voice announcing a call sign and possibly some other information, such as location or time. This is the repeater's *ID* and allows you to look up the call sign on-line and tell for sure where it is. The repeater may also send Morse code, a *CW ID*. This is very common and is a good reason to know Morse code, even if you don't use Morse to make contacts.

To find all of the repeaters in your area, you'll need a listing sorted by area, such as the *ARRL Repeater Directory*, or a source such as a club newsletter. Repeater frequencies are evenly spaced in *channels*, so you know exactly what frequency to use when you do find or select a repeater. If your radio has a *scanning* function, this would be a good time to learn how to use it. By continually scanning the repeater output frequencies, you'll eventually find all of the active local repeaters. Once you have located a frequency with an active repeater, you'll need to program your radio to use the correct offset and CTCSS tone. Offset is generally standardized on the different bands so that radios with an *autorepeat* function will automatically select the right amount and direction of frequency

144-148 MHz

Location	Output	Input	Call	Notes	Sponsor
ALABAMA					
Albertville	145.1100	–	KF4EYT	o(CA)e WXX 107.2	KF4EYT
Alexander City	146.9600	–	WA4KIK	oael 88.5	WA4KIK
Allsboro	147.0600	+	KB4FKU	o	KB4FKU
Andalusia	146.9400	–	WC4M	o 100.0	SARC
Andalusia	147.2600	+	WC4M	oewx 100.0	South AL R
Anniston	146.7800	–	KG4YRU	oaelwxz	Calhoun EM
Arab	146.9200	–	KE4Y	oaeRB WXZ 77.0	BMARA
Ashland	147.2550	+	KF4UOU	aewxz 131.8	KF4UOU
Athens	145.1500	–	N4SEV	o(CA)e WXZ 100.0/OR OFF/ARE	Limestone
Auburn	147.0600	–	KA4Y	o	KA4Y
Auburn	147.2400	+	K4RY	o	Auburn Uni
Bald Rock	145.1300	–	N4MLP	oerRB	St Clair C

Figure 4-3—Listings of repeaters, such as the ARRL's *Repeater Directory* or *TravelPlus for Repeaters* make it easy to find repeaters in your area or where you intend to travel. This is a typical *Repeater Directory* page.

Picking A Tone

As discussed in Section 3, CTCSS tones are added to your signal to cue the repeater that it should relay your signal. Before you can transmit through a repeater that requires a CTCSS tone, you'll have to find out which of the 38 possible tones it could be. If you have a repeater directory or are using a club Web site or newsletter, the tone will be listed along with the output frequency of the repeater. The listing will look like this:

PODUNK VALLEY   146.62 (-)   103.5

The (-) means that the repeater offset or shift is negative. You will hear the repeater output on 146.62 MHz and the input is 600 kHz (the standard 2-meter offset) below the output frequency, or 146.02 MHz. '103.5' is the frequency of the CTCSS or subaudible tone. Your radio's operating manual will explain how to select and activate the tone. There may be several tone options, such as tone squelch and digital code squelch (DCS). Leave them off for now.

Some radios also have the ability to determine the tone frequency from on-air signals. This is called *tone scan*. Radios with this feature can often be programmed to automatically set their own CTCSS tone to the same frequency without any operator intervention. This is very handy when you are new to an area, just visiting, or driving through.

shift. Choosing a subaudible tone is described in the sidebar.

Now you're ready to see if your signal is strong enough to be heard by the repeater and activate its transmitter. Activating the repeater is called *hitting the repeater* (or machine). Start by adjusting your squelch control so that the noise is just cut off. Remember to listen first to be sure you won't be interrupting any conversations, then press your PTT switch and say your call sign following by "Testing." Release the PTT switch and watch the signal strength indicator on your radio. If you were successful, the indicator display will show you that the repeater's output signal is being received. The repeater's output signal will be present for a few seconds and then you'll hear a "tsssssschht" sound. That sound is a *squelch tail*—the noise output by the repeater's receiver with no input signal before its own squelch circuit activates and shuts off audio output. The short delay in cutting off the transmitter is to keep the repeater from turning on and off rapidly due to a weak signal at the input.

ID and Control Topics

There you are, listening to a repeater conversation and suddenly…nothing. The repeater just shuts off—no output signal or anything! Has the repeater transmitter failed? Most likely, the repeater has *timed out* and stopping retransmitting the input signals. If you listen a little longer, the repeater will again respond and may even make a synthesized voice announcement, "Time out." One of the long-winded conversationalists will say, "We timed out the repeater…" and proceed.

Most repeaters start a timer when they begin transmitting. If the timer expires, typically in three minutes or so, without the transmitter turning off, the repeater turns off its transmitter. This prevents over-heating of the transmitter and gives stations a chance to break in by keeping one signal from occupying the repeater for long periods without a break. The timer resets when the transmitter turns off. i.e.—when the receiver does not detect an input signal. To reset the timer during your conversation, let the repeater *drop*. That is, stop transmitting long enough to hear the repeater's squelch tail as the transmitter shuts down.

Many repeaters are *linked* to other repeaters. That is, they share the audio signals each receives, retransmitting them over a wider area than any one repeater can cover. It is also common for repeaters to retransmit signals on other bands. For example, a 2-meter repeater linked to a 70-cm repeater allows stations on either band to contact each other. If the repeaters are *co-located*, meaning located at the same site, the repeaters can be physically connected with cables. Otherwise, a *control link* is required.

Control links consist of a transmitter and receiver that only relay audio and control signals between stations. They're not used for direct contacts and most are on the 1.25-meter and 70-cm bands. The signals carried by the links control various repeater features, usually enabling and disabling the retransmission of signals by the linked repeaters. Repeater *networks* or *systems* are made up of several linked repeaters that can be many miles apart. Control signals are used to configure the way in which the network relays signals between repeaters.

If you become a regular user of a linked repeater system, you may want to join the group operating the repeaters. You can be authorized to use the *control codes* (audio tones like those that dial telephones) to configure the repeater network and even perform basic maintenance and test functions. This is a valuable service you can provide that aids many hams.

Repeaters often have unexpected features, too. The repeater *controller* is a piece of equipment that regularly sends the repeater ID, operates the time-out timer, switches the transmitter on and off, etc. Modern controllers have microprocessors and sophisticated electronics that provide advanced features, such as synthesized voice announcements, time and date, weather conditions, and other interesting things. These are generally activated with control codes of their own.

To use control codes of any sort on the repeater requires that your transceiver be able to generate the appropriate tones or tone sequences. Check your radio's operating manual to see how that's done. It may be as simple as pressing the PTT switch and pressing the numeric keys on the radio. Be sure to identify your transmission by stating your call sign first. For example, "W1AW (tone) (tone) (tone)". Each repeater system will have its own protocol for using the control codes.

Repeater Etiquette

Repeaters are frequently like social clubs; they have a regular clientele and usually there are manners everyone uses to share the facility. To "join the club" you should learn and abide by those manners by listening. Here are some repeater manners shared by all repeaters users:

- Monitor before transmitting and keep transmissions short
- Identify your station legally; Identifying at the beginning of the contact is also a good idea
- Use no more power than is needed for a reliable contact
- Pause briefly between transmissions to listen for another station trying to break in

Remember, too, that using a repeater does not mean you are no longer responsible for operating legally. You are responsible, not the repeater owner, if you transmit communications that violate the FCC rules.

Giving signal reports on a repeater is also part of repeater etiquette. Because a repeater is re-transmitting the signal, it is not useful to report received signal strength. What's important is whether the transmitting station's signal is being received well by the repeater! Using the Q-system described earlier is common, often supplemented or replaced with the following terms:

- Full quieting—your signal is strong enough that no receiver noise is present
- White noise—not as strong as full quieting, some hiss is present
- Scratchy—weaker still, noise is almost as strong as your voice
- Mobile flutter or picket fencing—rapid fading due to moving through an area of marginal coverage
- Dropping out—mostly audible, but frequent periods of no signal
- Broken or breaking up—short periods of audibility, but mostly unreadable

These signal descriptions give a good picture of what the signal sounds like to the receiving station and can also give the transmitting station an idea about what to do, such as increase power or move to a better location. Users of handheld radios need to be continually aware of the need to keep their antenna vertical if their signal is less than full-quieting strength. Weak or *low* batteries are also a frequent problem for handheld users, so keep a fresh spare handy!

What if you receive a report that your signal's audio is strong, but distorted? A station that is slightly off-frequency will be strong, but distorted. This sometimes happens when a radio control key gets bumped, changing frequency by a small amount. If accidentally pressing a control key on your radio is a frequent problem (smaller radios are particularly prone to it), try using the LOCK feature of your radio to disable unintended key presses. Low batteries can also cause distorted audio. More common is the signal that sounds great for a couple of seconds at first, then suddenly fades. The batteries are too weak to sustain a transmission and the transmitter shuts off. Time for a fresh battery pack!

Autopatch

With the advent of widespread mobile phone systems, a connection to the telephone system through a repeater has less importance than it once did. Nevertheless, repeaters often have coverage where mobile phone systems don't and mobile phone batteries do occasionally discharge! The system that allows hams to make a phone call through a repeater is called *autopatch*, short for "automatic phone patch." ("Patch" is slang for a connection.)

When activated by the proper control code sequence, the autopatch system connects the repeater's transmitter and receiver to the phone line and you'll hear dial tone over the air. The ham then dials the number by pressing the PTT switch and pressing numeric keys. From there, it's just like a regular telephone call except that you have to press the PTT switch to talk and you can't talk and listen at the same time.

When using the autopatch, follow any repeater rules about using the autopatch system; frequency of use, where you can call, how long calls may be, etc. You are also responsible for following all FCC rules about allowable transmissions. As soon as the called party picks up the phone, you should identify the call as being via Amateur Radio to avoid confusion and prohibited transmissions. You are also required to abide by all identification rules while placing and conducting the call, as well.

Open, Special Use, and Private Repeaters

If you look in a repeater directory, you may see a symbol in the listing that indicates a repeater is *closed*. That means the repeater is not available for public use—only authorized stations may use the repeater. Other repeaters are dedicated to a special purpose, such as emergency communications. In both cases, the repeater owners prefer to restrict the use of their repeater. This is perfectly legitimate and within FCC rules.

Most repeaters are open and free for anyone to use. Closed repeaters usually require membership in a group that supports the expenses of operating and maintaining a repeater—it may have special features or capabilities. To find out how to join the group, enter the repeater's call sign into the "Call Sign Search" window on the ARRL's home page at **www.arrl.org**. Contact the repeater group via the mailing address shown.

REPEATER COORDINATION

It's a natural question, "Who decides what repeater can use a specific pair of frequencies?" You may be surprised to learn that hams decide themselves and the FCC has nothing to do with it. This is part of the Amateur Radio tradition of self-policing and self-administration. Hams developed the system of regional frequency coordinators to insure that repeaters use the amateur bands wisely and avoid interference to the greatest degree possible. Each region of the US has its own frequency coordinator, made up of a team of volunteers. Where regions overlap, the coordinators work together to minimize interference and keep the coordination process orderly. A list of frequency coordinators is available on the Web site of the National Council of Frequency Coordinators at **www.arrl.org/nfcc/coordinators.htm**.

A *coordinated* repeater uses frequencies approved by a regional coordinator. *Uncoordinated* repeaters are strongly discouraged because they often cause interference. Before putting their repeater on the air, repeater owners apply to their region's coordinator for an available pair of frequencies. The coordinators determine what frequencies are best suited for the repeater's location. Once the frequency pair is assigned, the repeater owner can then turn on their repeater.

SIMPLEX CHANNELS

Repeaters provide such good coverage, why would anyone not use a repeater for a contact? A repeater's excellent coverage and signal strength are precisely why it's not always appropriate to use them. Only so many repeaters can share a band in a region so hams must use them wisely. It's easy to become so used to using repeaters that direct or *simplex* communication isn't considered. In fact, it's often quite easy to make contact directly from point-to-point.

If you are close to the station with whom you're in contact, why not give simplex a try? It avoids tying up a repeater and makes your conversation a lot less public than having it broadcast over the repeater's entire coverage area! Here's how to make a move to a frequency for simplex communications:

W1AW: "NK7U this is W1AW, are you on the repeater this morning?"

NK7U: "W1AW this is NK7U. Yes, and you're strong on the input. Let's move to 146.55 simplex."

W1AW: "OK, I'll meet you on 146.55. W1AW clear."

NK7U: "NK7U clear"

You can tell if you are within range just by listening to the other station transmit on the repeater's input frequency. Many radios have a "reverse" function that swaps your transmit and receive frequencies, making it easy to listen for the other station.

Listening on the repeater's input is often helpful when a weak station is trying to access the repeater, but isn't quite strong enough. If the weak station is near you, it's likely that you'll hear their signal better than the repeater does!

Simplex channels are conveniently located between bands of repeater input and output channels. For example, on 2-meters, simplex channels are found from 146.42 to 146.58 MHz and from 147.42 to 147.58 MHz. This means the antenna you use for repeater contacts will work just fine for simplex, too.

REPEATER—DIGITAL SYSTEMS

Ham radio and the Internet each have complementary advantages. Hams can roam freely, using repeaters from a vehicle, at home, or on foot. Then Internet provides a high-speed connection between two points—nearly anywhere on Earth. It's a natural to combine the two and several systems do just that.

- IRLP (Internet Radio Linking Project)—**www.irlp.org**
- Echolink—**www.echolink.org**
- WIRES II—**www.vxstd.com/en/wiresinfo-en**—a proprietary system of the Yaesu company
- D-STAR—**www.icomamerica.com/amateur/dstar**—a system based on the public D-STAR standard

The two most popular systems, IRLP and Echolink use VoIP (Voice over Internet Protocol) technology to link repeaters. It's the same technology used by Internet tele-

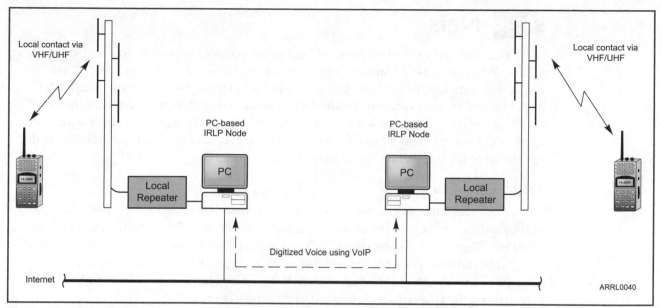

Local contact via VHF/UHF

PC-based IRLP Node

PC

Local Repeater

PC-based IRLP Node

PC

Local Repeater

Local contact via VHF/UHF

Digitized Voice using VoIP

Internet

ARRL0040

Figure 4-4—IRLP and Echolink are systems of repeaters linked by the Internet protocol VoIP. Hams can use a local repeater and radio to make contacts worldwide by using control codes to connect to far-away repeaters.

phone service vendors, such as Skype. The main difference between IRLP and Echolink is that IRLP requires all audio to be transmitted into the system via a radio link. (IRLP does allow a PC user to *listen* to conversations.) That means you must be a licensed amateur to use repeaters linked by IRLP. Echolink allows audio to come from a PC and microphone, so a radio is not necessary but hams are required to send a copy of their license to the Echolink system administrators to be authorized to use the system.

The D-STAR system provides voice communications and adds the ability to send data across the system, as well. D-STAR is closer to an Ethernet network and in its high-speed form, allows a computer with a Web browser to be connected to the Internet via ham radio. The D-STAR standard was developed by the Japanese Amateur Radio League (JARL) and is currently implemented in equipment manufactured by the ICOM company. Wires II uses a proprietary standard developed by the Yaesu company and is a voice-only system. More protocols and systems are likely to appear in the coming years, so keep an eye on newsletters and magazines for information about them.

How does an IRLP or Echolink contact differ from a regular repeater contact? To initiate an IRLP or Echolink contact, the initiating station must know the repeater control code to request an IRLP connection—this is the ON code. The ON code varies from repeater to repeater and may require membership in a club. Once the ON code is entered via your keypad, the 4-digit code for the IRLP *node*—a destination repeater—is entered. You will hear a confirmation tone or an error tone. If confirmed, announce your presence on the destination repeater as if you were operating locally. IRLP nodes are listed in directories and on the Internet.

At the destination repeater, operation is more like a regular repeater contact. You will hear a tone as the remote station connects via the IRLP network, then the connecting station will be heard. It can be disconcerting to be driving to work and hear an Australian or Russian voice on your local repeater! Nevertheless, it's fun to meet hams from around the world. Much more information on Internet-linked systems is available in the ARRL book *VoIP: Internet Linking for Radio Amateurs*.

Study questions T3A01-04, 08-11; T3B04-06; T3C01-03; T3D06; T5C02, 04, 09-14. T5D12; T6B01-11; T6C08-09. Review this section if you have difficulty.

4.4 Nets

As you read and experience more of ham radio, you will encounter frequent references to *nets*. What are these mysterious organizations and how can you participate? "Net" is just an abbreviation for *network*. Developed in the very early days of radio, these networks helped stations meet on the air to share news and exchange messages, called *traffic* in net lingo. The modern network, now associated with computers, is a direct descendent of the radio net and uses many of the same terms and concepts. There are lots of nets that are available for the Technician-class licensee.

TYPES OF NETS

Amateur radio makes extensive use of nets and there are three major types; social, traffic, and emergency. Nets can serve a group as small as a club or be international in coverage. Some are very formal and follow a strict procedure while others are more like a group conversation. Each net has a theme or purpose and has a regular schedule to convene at specific frequencies and times. Nets can be found on both HF and VHF/UHF bands, using the frequencies best suited to their coverage needs.

Social

The least formal and most common are the social nets, ham radio's on-the-air meetings. Themes of these nets varies widely, from hobbies (stamp collecting, model rocketry, chess, etc.) to award chasing (DX and County Hunting) to "stay-in-contact" nets that support RVers and boating stations as they travel. These nets are easy to join and rarely require any training or special procedures. Just listen to how the net conducts its business and try to emulate it.

Traffic

This is the "original net"—an on-the-air method of exchanging and routing messages, known as *radiograms*, across town or across the country. An extensive structure of traffic nets, the National Traffic System (NTS) is designed for efficient station-to-station *traffic handling*. The NTS is composed of local, regional, and national nets through which the radiograms pass, station-to-station, until a local net member passes them to the recipient. The NTS is active every day, especially during emergency and disaster recoveries, such as from hurricanes. Traffic nets follow a set procedure that can be quickly learned. Local VHF traffic nets are easy to join and welcome newcomers.

Emergency & Public Service

Hams coordinate their emergency response activities with nets that spring into action whenever they're needed. In areas where severe weather is common, nets monitor weather conditions before a storm and assist with recovery in case damage results. Emergency nets are frequently dual-purpose; they pass emergency traffic and coordinate reporting and response activities. While waiting to be called to service, emergency nets hold regular information and training sessions. That is the time to join the group and learn the procedures, not after disaster strikes!

As a way to practice skills and provide service at the same time, hams also provide communications for public events, such as festivals, parades, and sporting events. These public service activities are organized on the ground and on the air just as an emergency response would require. You can volunteer for one of these events and that is how many hams developed into skilled emergency response operators!

NET STRUCTURE & PARTICIPATION

In order to participate effectively in a net, you need to know a little bit about how a net is organized. Knowing the common procedures and signals is also important. Knowing a few of the rules of the road keeps traffic flowing smoothly, both for vehicles and nets. (There is a great deal of information about nets and public service on the ARRL Web site at **www.arrl.org/hrlm/section-4.html**.)

Net structure centers on the *net control station* (NCS). The NCS is responsible for conducting each session of the net in an orderly way that follows established net procedures. It is important that the NCS has a strong and clear signal that can be heard clearly by all net members, whether the signal is relayed through a repeater or transmitted directly on SSB or CW. The net control position should be filled by a skilled and experienced operator.

A typical net session begins with the NCS calling the net to order by reading an opening *script* that establishes the purpose of the net. The next step is generally to see if any stations have emergency needs or messages and address them promptly. Announcements of interest to all net members are then read and finally, individual members *check in* at the direction of the NCS. With all in order, the net's business is then conducted, whether it is a swap-and-shop session or one that handles messages from a Red Cross shelter. **Table 4-5** lists a number of the common Q-signals and procedural signals and definitions used in net operation.

Table 4-5
Common Net Q-Signals and Procedural Signals

QNI	Net stations report in
QNU	The net has traffic for you
QRU	I have nothing for you
QRV	I am ready to copy
QSP	Please relay the following information
QTC	I have the following traffic
$\overline{AR}$	End of message
$\overline{AS}$	Stand by
C	Yes
GG	Going (as in changing frequency)

Emergency Nets

There are some important differences in operations when a net is responding to a disaster or emergency. Efficiency and accuracy become the highest priorities so that the most important business is handled at all times. This requires all stations to follow *net discipline*, following directions and net procedures at all times. Does this mean you must be an expert operator to participate in emergency nets? No, but you must be willing to listen and to follow directions!

Listening is a very important skill under these conditions. It is natural to want to jump in with comments and suggestions under stressful circumstances, but doing so often causes delays and mistakes. Once you have checked into an emergency net, you should not transmit unless you are specifically requested or authorized to do so or a request in made for capabilities or information that you can provide.

Once an emergency net is established, the NCS must continue to make emergency communications the highest priority at all times. This includes emergency messages, such as radiograms with the Emergency precedence. Even if other messages have been waiting, emergency traffic of any sort has the highest priority. The order of priority for communications is Emergency, Priority, Health-and-Welfare, then Routine.

What happens if you are the first station to show up on the net frequency? Even though you may not be a skilled NCS, you are encouraged to take the initiative and begin the net on your own. If you have the net's script on hand, follow it. If not, announce that you are acting as a temporary net control and make a list of members as they check in including their location and whether they have messages or information. When a more experienced station is available, you can hand off net control duties and read the check-in list.

The American Radio Relay League
RADIOGRAM
Via Amateur Radio

Number	Precedence	HX	Station of Origin	Check	Place of Origin	Time Filed	Date
207	P	E	W1FN	10	LEBANON NH	1200 EST	JAN 4

To:
MARK DOE
RED CROSS DISASTER OFFICE
123 MAIN ST
RUTLAND VT 05701
Telephone Number: 802-555-1212

This Radio Message was received at:
Amateur Station_____ Date_____
Name_____
Street Address_____
City, State, Zip_____

NEED MORE COTS AND SANITATION
KITS AT ALL FIVE SHELTERS

_____ _____ _____ _____
_____ _____ JOAN SMITH SHELTER MANAGER

	From	Date	Time		To	Date	Time
REC'D				SENT			

A licensed Amateur Radio Operator, whose address is shown above, handled this message free of charge. As such messages are handled solely for the pleasure of operating, a 'Ham' Operator can accept no compensation. A return message may be filed with the 'Ham' delivering this message to you. Further information on Amateur Radio may be obtained from ARRL Headquarters, 225, Main Street, Newington, CT 06111.

The American Radio Relay League, Inc. is the National Membership Society of licensed radio amateurs and the publisher of QST Magazine. One of its functions is promotion of public service communication among Amateur Operators. To that end, The League has organized the National Traffic System for daily nationwide message handling.

Figure 4-5—The ARRL Radiogram form is the standard for originating and relaying messages. The preamble identifies the message and allows it to be tracked. The limited text requires that the message be focused, balancing length against minimizing errors during transmission.

Check and Balances

The radiogram, while old, has its own origins in the telegraph message, developed in the middle of the 1800s. Telegraph companies quickly learned that it was important to be able to track messages and minimize relaying errors. They developed the idea of a *preamble* that contained the information describing the message and providing the means to trace it back to its origins.

The preamble concept proved so useful that it has been carried forward into the terminology of the Internet and computer networks. For example, "preamble" is still used in the very same way—the information at the head of a transmission containing information about the message that follows. If you examine a transmission of data on a garden-variety Ethernet network, you will find that each packet of data has a preamble with a unique number, and address, among other things.

Internet messages are also protected against error by the use of *checksums*, a word derived from the radiogram's *check*. Both serve as error-detection mechanisms. If a radiogram is received with a check different than the number of words in the text or if a network packet is received with a checksum that doesn't match what the receiver thinks it should be, then the information is not correct and retransmission is required.

Everywhere you look in modern network technology, you will find the echoes of the landline telegraphers and radio operators!

TRAFFIC HANDLING

An important part of emergency and disaster net operation is the ability to accurately transfer information as formal messages called radiograms. A blank radiogram form is shown in **Figure 4-5**—you can download your own radiogram from at **www.arrl.org/radiogram**. A radiogram has three parts; the preamble (including the address), the body, and the signature.

The preamble is made up of several bits of information about the message. These establish a unique identity for each message so that it can be handled and tracked appropriately as it moves through the Amateur Radio traffic handling system.

- Number—a unique number assigned by the station that creates the radiogram
- Precedence—a description of the nature of the radiogram; Routine, Priority, Emergency, and Welfare
- Handling Instructions (HX)—for special instructions in how to the handle the radiogram.
- Station of Origin—the call sign of the radio station from which the radiogram was first sent by Amateur Radio. (This allows information about the message to be returned to the sending station.)
- Check—the number of words in the text of the radiogram.
- Place of Origin—is the name of the town from which the radiogram started.
- Time and Date—is the time and date the radiogram is received at the station that first sent it.
- Address—the complete name, street and number, city and state to whom the radiogram is going

Following the preamble comes the text of the radiogram. The ARRL radiogram limits the text to 25 total words. That doesn't sound like a lot, but the limit helps to focus the message topic and eliminate unnecessary words. Longer messages also are more difficult to relay without errors, so there must be a balance between volume and accuracy. The signature which follows the text is usually the name of the person originating the message.

If you are asked to generate a radiogram,

you should follow this format. If you can not decide on what to use for some of the preamble contents, leave them blank. In an emergency, you may not always have all of the information. Another net station may be able to help you or the message can be sent without the information. In an emergency, though, the one item that must be included is the name of the person originating the message.

If you check in to a traffic handling net, you will find that they make heavy use of the Q-signals you learned about earlier; their clear meaning and short length greatly speed up operations. They can be found in the *ARRL Operating Manual* which includes a handy reference on traffic handling. The radiogram quick-reference card, ARRL FSD-218 can also be downloaded from **www.arrl.org/FSD-218**.

FINDING NETS

You can find out what nets operate in your area by using the on-line ARRL Net Directory at **www.arrl.org/nets**. Click on "Net Directory Search" to access the on-line search page. To get an idea of the nets in your area, select "State Nets" and select your state from the list. You'll be surprised at how many there are to choose from! Click on the name of the net for detailed information about the net including an e-mail address for the net's manager. A printed version of the directory is also available for use in the field.

> *Study questions T8C01, 03, 04, 06-11.*
> *Review this section if you have difficulty.*

4.5 | Emergency Operating

Providing communications assistance during an emergency or in response to a natural disaster is one of the Amateur Service's most important reasons for existing at all. In fact, in Part 97.1, the Basis and Purpose for the Amateur Service, emergency communications is the very first reason! Many hams are regularly involved with emergency communications or *emcomm*. It may be your reason for becoming interested in ham radio. Luckily, emcomm is not greatly different than what hams do on a day-to-day basis so you won't have to completely re-learn your skills. In this part of Section 4, we'll cover the important differences.

OPERATING UNDER EMERGENCY CONDITIONS OR IN PUBLIC SERVICE

As a licensed communicator, you are encouraged to get involved during emergency conditions. Use what you have learned! The FCC places top priority on emergency communications which have priority over all other types of Amateur Radio communications on any frequency. The FCC also recognizes the need for flexibility under such circumstances. For example, in defining emergency communications, Section 97.401(a) says:

"When normal communication systems are overloaded, damaged or disrupted because a disaster has occurred, or is likely to occur...an amateur station may make transmissions necessary to meet essential communication needs and facilitate relief actions."

If you're in the middle of a hurricane, tornado, or blizzard, and you offer your communications services to the authorities, you are permitted to do whatever you need to do to help deal with the emergency. Public safety or medical personnel can use your radio. You can work for the Red Cross to perform damage assessment. In an emergency situation where there is immediate risk to life or property, you may use any means possible to address that risk. You are only prohibited from transmitting information in support of a commercial business or confidential personal information of a third party such as a disaster victim without their consent.

Similarly, in an emergency situation you may use whatever communications means is at hand to respond—any means on any frequency. If a fire department radio or marine SSB transceiver is all that's available, by all means use it to call any station you think might hear you! This waiver of normal rules lasts as long as the threat to life and property remains imminent. Once the threat has receded, you must return to normal rules, even in support of public safety agencies. At all times, you are bound by FCC rules, even if using your radio in support of a public safety agency. For example, while providing post-event communications at a fire department command post, you are not permitted to use a modified ham radio on fire department frequencies.

News messages and reports are not considered emergency communications by the FCC. You are not allowed to relay such information via Amateur Radio. Inform reporters that ask you to relay reports that you can't do that under FCC rules.

While you are providing emcomm, you must also remember to operate efficiently and strive for the highest levels of performance. Here are some tips for good emergency operating;

- Don't become part of the problem—you are there to assist, not become a victim or act as a first responder.
- Never speculate or guess—strive for 100% accuracy and don't be afraid to say "I don't know." Rumors are impossible to stop, once started.
- Don't give out unauthorized information—reporters and members of the public are often desperate for information. Direct them to the appropriate spokesperson or information source.

- Maintain Your Safety—you are no help if you become injured or put yourself at risk
- Maintain Radio Discipline—stay with the established protocols and refrain from idle conversation on emergency frequencies that might impede emergency communications
- Protect personal information—never send confidential personal information via Amateur Radio without consent. If possible, use modes such as packet radio or Morse code that are less likely to be overheard by the public.

EMERGENCY DECLARATIONS

In a serious, widespread emergency, the FCC may declare a *temporary state of communications emergency*. The declaration will contain any special conditions or rules that are to be observed during the emergency. The declaration is in force until the FCC lifts it. Communications emergency declarations are distributed by the FCC through its Web site, via ARRL bulletins on headquarters station W1AW and the ARRL Web site, and through the National Traffic System and Official Relay Stations. Amateur Web sites and e-mail lists pick up the declarations and relay them throughout the amateur community.

Normal rules may be suspended or modified and frequencies may be reserved exclusively for emergency communications. An emergency declaration may authorize amateurs to communicate with stations in other services, for example. More commonly, certain frequencies will be reserved for emergency communications. You must then avoid those frequencies unless you are participating in the relief effort. A declared communications emergency is the only time that a frequency is legally restricted to one use. Otherwise, no station has the exclusive use of any frequency although hams are expected to be cooperative and assist with emergency operations.

DISTRESS CALLS

If you are in immediate danger or require immediate emergency help, you may make a distress call on any frequency you feel offers you a chance of being heard. In these circumstances, here's what to do:
- On a voice mode, say "Mayday Mayday Mayday" or on CW send "SOS SOS SOS" (Mayday should not be confused with the Pan-Pan urgency call) followed by "any station come in please"
- Identify the transmission with your call sign.
- Give your location with enough detail to be located
- State the nature of the situation
- Describe the type of assistance required
- Give any other pertinent information

Then pause for any station to answer. Repeat the procedure as long as possible or until you get an answer. The reason for giving all of the information during each call is so that if you can't hear responding stations, they'll still learn where you are and what help you need. Under no circumstances make a false distress call or allow others to do so using your equipment. Your amateur license could be revoked and you could be subject to a substantial penalty or even imprisonment.

If you hear a distress call—on any frequency—you may respond. Outside the amateur bands, such as on the international marine distress calling frequency of 2182 kHz, be sure that no other station or vessel is responding before you call the station. Inside the amateur bands, suspend any other ongoing communications immediately. Record everything the station sends and then respond. If they hear you, let them know that you have copied their information, clarify any information as required and immediately contact the proper authorities. Stay on frequency with the station in distress until authorities are either on frequency or arrive at the scene.

TACTICAL COMMUNICATIONS

Tactical communications is performed between emergency responders or public service providers involving a few people in a small area, such as a responding to a fire or landslide. The communications serve to coordinate activities ("Go to the south parking lot"), report status ("The final float is leaving the staging area"), or request resources ("First aid is needed at 232nd and the highway."). This type of message is rarely recorded and is not passed in radiogram format.

Because of its short range, tactical needs are usually satisfied by using VHF/UHF simplex or repeater channels. Mobile, portable, and handheld radios are particularly useful when working with public safety and government agencies.

To increase efficiency and smooth coordination, stations engaged in tactical communications should use *tactical call signs*, such as "Command Post Three" or "School Kitchen" or "Judges Stand". These describe a function, location, or organization. This allows operators to change without changing the call sign of the stations and frees non-amateur personnel from having to use amateur call signs.

Tactical call signs do not, however, satisfy the FCC regulations for station identification, which must be followed. To properly identify, each operator should identify with both their personal call sign and the tactical call sign at the beginning and end of their operation and at least once every 10 minutes during their shift of operation.

EMERGENCY EQUIPMENT

When a disaster strikes or emergency occurs, the need for communications often can't be filled from home stations. In cases like these, hams use portable and mobile stations

Figure 4-6—A "go-kit" contains the essentials for quick response to an emergency that may keep you in the field for up to 24 hours. It should contain both personal and radio items so that you can be effective without requiring additional support.

and handheld radios that operate without the need for commercial power. To be effective and respond quickly, you should have a "go-kit" made up that can support you for at least 24 hours. **Figure 4-6** shows a typical go-kit. The Web sites at **www.arrl.org/hrlm/section-4.html** have more information about emergency preparedness. If you are part of an emcomm team, they will likely have a go-kit list already prepared and tested.

Alternate power sources will likely be required in any serious emergency or response to a rural area. Mobile and portable stations can often use a car battery as a power source, recharging it from any available vehicle with jumper cables. For true energy independence a solar-powered system or even a bicycle generator is used.

A handheld transceiver is quite useful, but needs some support to be effective for a day of use. Start with spare battery packs and keep them charged. Many radios have battery packs that accept alkaline AA cells. These may be more useful in an emergency because they don't require a charger and AA batteries are widely available. Have an external antenna, such as mobile mag-mount whip or even a wire dipole to improve the signal from your handheld. A power adapter that can plug into a vehicle's cigar lighter or attach to a car battery is also quite useful. Take along a list of repeaters and simplex frequencies for quick reference and know how to use your radio!

Before responding, think about the special circumstances you are likely to encounter. Are you likely to be in a noisy place? A headset with earphones might be important. What about the weather and temperature? Be sure you have adequate clothing and protection for your anticipated situation.

RACES AND ARES

The two largest Amateur Radio emergency response organizations are ARES® (Amateur Radio Emergency Service) sponsored by the ARRL and the Radio Amateur Civil Emergency Service (RACES). ARES members support local and regional government and non-governmental agencies such as the Red Cross, Salvation Army, and National Weather Service during emergencies. Any licensed amateur can participate in ARES. RACES is a special part of the Amateur Service created by the FCC. Amateurs that register with the local, state, or federal government emergency management agencies provide communications assistance during civil emergencies. (See Part 97.407 of the FCC Rules for more information on RACES.) Many amateurs are members of both ARES and RACES teams so that they can respond to either need.

NIMS AND FEMA

Emergency Communications Training

To be truly effective when responding to an emergency, you need some training and even better, some practice opportunities! Doesn't music sound a lot better when the musicians have learned the music and practiced it before the show?

Start by joining a local amateur emergency preparedness team. Your local radio club, ARES team, RACES team, county Search-and-Rescue, or Salvation Army chapter (just to name a few) are all organizations to investigate. Choose the one that suits your interests.

Take advantage of any training the group might provide or recommend. For example, the ARRL offers three levels of Emergency Communications on-line training classes. (See **www.arrl.org/cce** for more information on EC-001, the Level 1 Amateur Radio Emergency Communications course.) The Federal Emergency Management Agency (FEMA) offers free emergency preparedness training course at their Web site **www.fema.gov**. The courses on the National Incident Management System (NIMS) are very helpful to learn how public safety agencies will be organized in a disaster.

When your group has a drill or exercise, try to participate, even if just in the planning and organization stage. The experience will serve you well when a real activation occurs.

ARRL Field Day—The Biggest Amateur Event of All!

Every year on the last full weekend of June, North American hams head for the hills…and the fields and the parks and the backyards. It's Field Day! This is the annual emergency-preparedness exercise in which more hams participate than any other. The basic idea—set up a portable station (or several) and try to make as many contacts with other ham groups as possible on as many amateur bands as possible. If you think the bands are busy on weekends, wait until you hear them on ARRL Field Day!

Some groups focus on the emergency preparedness aspect, others get into the competitive aspect trying for the most points, while some just treat it as the annual club picnic plus radio operating. Whatever your organization prefers, Field Day is a great way to see a lot of ham radio all in one spot and all at the same time. For more information, read the Field Day announcement in the May issue of *QST* magazine or the event summary and results in the December issue. CQ, Field Day!

www.arrl.org/FieldDay

The ARRL sponsors an annual Simulated Emergency Test in October of every year. If your club participates in ARRL Field Day (see sidebar), be sure to attend. Almost every town has at least one public event with communications needs that Amateur Radio could fill. Help plan to put an amateur team on the job!

Finally, test your own preparedness! Check your go-kit and emergency equipment every six months to be sure it's all together and working. Double check your power sources, especially batteries that might grow weak over time. You're only as effective as your equipment will let you be.

Study questions T3D09; T3D10; T7A01,02,04; T8A01-12; T8B01-11; T8C02,05; T0A09. Review this section if you have difficulty.

4.6 Special Modes and Techniques

We've just scratched the surface of ham operating styles and opportunities. The longer you stay with the hobby, the more different parts of it you discover! Here is a survey of some activities that you are likely to encounter as you enter the hobby.

DXING, AWARDS, & CONTESTING

Since the beginning of radio, even before amateurs appeared on the scene, operators strove to make contact over longer and longer distances. Marconi himself started by sending messages across a few hundred yards and gradually built up his capabilities to where he could span the Atlantic Ocean. Amateur radio is no different. An enduring and popular past time is to see if you can pull in far away signals from away over the horizon, exchanging QSL cards to confirm and remember the contact.

This is called DXing, where DX stands for "distant station". Distance is a relative thing. DX means thousands of miles on HF and occasionally 6 meters. At VHF/UHF, any contact beyond the radio horizon is considered DX. Microwave operators scout out locations with unobstructed views to make contacts of many miles. Making DX contacts is best done on SSB or CW because of the efficiency of those modes. The articles and links at **www.arrl.org/tis/info/propagation.html** will help you learn about propagation on the different bands.

On HF, simple antennas, such as dipoles and verticals, are sufficient when the band is open for long-distance propagation. The key is to learn when the best times are for propagation to various areas around the world and be on the band at that time. Start by responding to a DX station calling CQ. If your signal is strong, the contact may turn into a short ragchew. If signals are weak, your contact may be limited to just an exchange of signal report, location, and name. Log each contact using UTC time (the 24-hour time broadcast by WWV on 2.5, 5, 10 and 15 MHz) and send your QSL card direct to the DX station (be sure to include return postage) or via the ARRL QSL Bureau.

VHF/UHF DX contacts tend to be much shorter, since the time during which the band opens for long-distance propagation is usually brief. The important piece of information to exchange is your grid square, discussed earlier in this section in the paragraph on locators. While a dipole will work at times, you'll get much better results by using a beam antenna that you can point in different directions. Your antennas should be horizontally polarized—the norm for VHF/UHF SSB and CW. Again, take care to log your contacts properly with an accurate time.

Pursuing long-distance contacts really hones a ham's technical and operating skills. In the course of DXing, one learns many things about propagation, antennas, and the natural environment! To recognize the achievements of DXers there are many awards offered by the ARRL and other organizations. For contacting the numerous countries the DXCC award

Figure 4-7—The 1000 Miles Per Watt award recognizes a low-power station's ability to be heard at long distances. This is just one of thousands of operating awards available to hams for every imaginable type of operation achievement.

is popular. VHF/UHF enthusiasts contact grid squares for the VUCC award. Contacting all of the US states (Worked All States) is popular around the world.

Aside from DXing, there are many awards for which a ham can qualify in the course of casual operating. Many radio clubs offer awards for contacting their members, for example. Low power enthusiasts, such as the QRP ARCI offer the award shown in **Figure 4-7** for making a contact that spans 1000 miles per watt of power. The list of awards is staggering. Ted K1BV compiles an annual awards directory that lists over 3,300 awards! (**www.dxawards.com/book.html**). Do you think you could qualify for the Tasmanian Devil Award?

Focusing the competitive urge even more, radio *contests* are held in which the competitors try to make as many short contacts in a fixed period of time as possible. Some contests are very short, called *sprints*, while others last an entire weekend. There are contests that use just one band or mode and others that span multiple bands and multiple modes. If you encounter a contest on the air, jump in and make a few contacts. You'll be asked for some information called an *exchange*. It may consist of your location, a signal report, and a *serial number* that's the number of contacts you've made in the contest so far. Just ask, "What do you need?" The contest station will help you provide the right information. You'll find that it's a lot of fun for even a casual operator!

As you might imagine, consistent efforts of this sort make for a very capable operator. The excellent stations and skills of contest operators are quite applicable to emergency operating and traffic handling, as well. In fact, many of today's top contest operators got their start in handling radiograms and net operations! The ARRL and other organizations sponsor contests that run the gamut from international events attracting thousands to quiet, relaxed competitions to contact lighthouses or islands. You can find the rules for these events on the ARRL's *Contest Corral* Web page at **www.arrl.org/contests**.

Special Events

In between the contesters and the DXers are the *special event* stations that operate for a short period to commemorate or publicize an activity of special significance. For example, a club might set up a station at a state fair or a sporting event. Each December, the W1AA club operates from the site of Marconi's Cape Cod station.

Aside from the sheer novelty of it, the stations often offer a unique or colorful QSL card or certificate for contacting them. Many of them also obtain special call signs that can only be logged during their activity. You'll find these stations on the air from all around the world and collecting their QSL cards is a popular past time for many amateurs. Contacts with special event stations are conducted much like DX contacts.

Over the River and Through the Woods

A different and more physical type of contest is known as *foxhunting*. Locating a hidden transmitter (the fox) has been a popular ham activity for many years. It has its practical side, too, training hams to find downed aircraft, lost hikers, and sources of interference. You don't need much in the way of equipment; a portable radio with a signal strength indicator and a handheld or portable directional antenna, such as a small Yagi beam. One ham hides the transmitter (hams can be very devious and inventive when it comes to hiding places) and the rest drive, walk, or bike the area taking bearings and attempting to be the first to locate the transmitter.

In recent years a new type of outdoor radiosport has reached US shores from Europe and Asia— *direction finding*. Held as organized events, direction finding is a hybrid of the radio fox hunt and orienteering skills to navigate outdoors with map and compass. The US Amateur Radio Direction Finding organization (**www.ardfusa.com**) is just one of a number of national groups in this worldwide sport, especially popular with teens and young adults. If you are a hiker or camper, then you might be interested in applying your outdoor skills to ARDF.

SATELLITES

Communicating through an amateur satellite sounds like a very complicated, high-tech effort, but it can be quite simple. What you need is a radio that can transmit on one band and listen on another (most can, even handhelds) and a way to find out when the satellite is in view. A satellite far enough above the Earth can even relay signals between countries. Some amateurs have even obtained the coveted DXCC award for contacting 100 different countries through satellites!

Amateurs have built satellites since 1961, launching them when extra space is available in a commercial payload. Amateur satellites are nicknamed *OSCAR* for *Orbiting Satellite Carrying Amateur Radio*. They relay signals between VHF and UHF bands. (A few satellites act as FM repeaters and can be accessed with regular FM rigs.) The ionosphere is usually transparent to signals at these frequencies, so the signals can pass between earth and space easily. If two stations both have the satellite in view at the same time as shown in **Figure 4-8**, they can make contact via line-of-sight propagation.

Satellite contacts can be made by any amateur licensed to transmit on the *uplink* frequency. For example, a Technician-class licensee could use a satellite that is listening for uplink signals on 2 meters and transmitting on a 10-meter *downlink* frequency even though not permitted to transmit on 10-meters. Satellite uplink and downlink frequencies are restricted to the special satellite *sub-bands* listed in **Table 4-6**, segments of frequencies set aside for earth-space communications.

- Apogee—The point of a satellite's orbit that is farthest from Earth
- Beacon—A signal broadcast continuously that contains information about a satellite
- Doppler shift—A shift in a signal's frequency due to relative motion between the transmitter and receiver
- Perigee—The point of a satellite's orbit that is nearest the Earth
- Keplerian elements—A set of numbers that describe the satellite's orbit so that it can be tracked
- LEO—Low Earth Orbit
- Elliptical orbit—An orbit with a large difference between apogee and perigee

To find out when the satellite will be in view, you'll need a *satellite tracking program*. You will also need to enter the Keplerian elements for the satellite. It is a good practice technique to enter the elements for the International Space Station (ISS) and then find it visually in the sky, watching it pass overhead at sunset or dawn. The ISS also carries an amateur station that you can contact by packet, or if you're lucky, one of the astronauts by voice! You can also practice by listening for the beacons of satellites as they pass by.

When you are ready to try a satellite contact, known as *squirting the bird*, you'll get best results with a beam antenna that you can aim at the

Table 4-6

Satellite Sub-bands

29.300-29.510 MHz

144.30-144.50 and 145.80-146.00 MHz

435.00-438.00 MHz

1260-1270 MHz

2400-2410 and 2430-2438 MHz

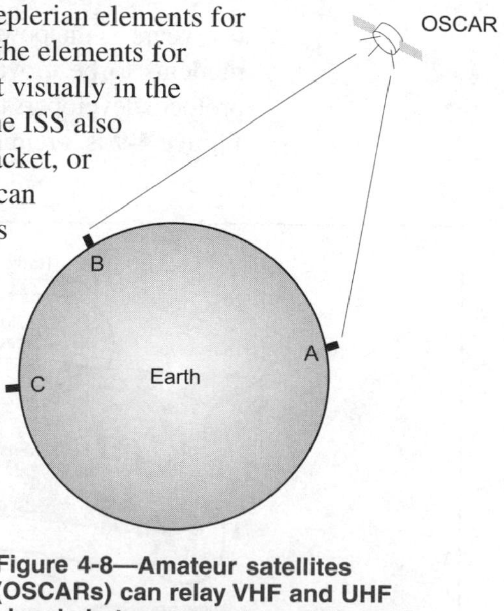

Figure 4-8—Amateur satellites (OSCARs) can relay VHF and UHF signals between any two stations that both have the satellite in view at the same time. Contacts through the satellite are possible between station A and B, but not with station C.

Working the International Space Station

It is a pleasant surprise to learn that Amateur Radio has a place on the International Space Station (ISS)! Not only that, but nearly all of the astronauts hold Amateur Radio licenses and some are quite active from their orbiting home on the ISS. Any amateur licensed to use the 2-meter band can join in the fun from Earth.

The astronauts frequently operate using FM voice. A packet bulletin board system (BBS) and a voice FM system are both on-board. One or the other is active at all times—you can check the mode of the station on the AMSAT Web site. Recently, the ISS station also added an APRS (Automatic Position Reporting System) digipeater!

When the ISS is in view, you can connect to the packet BBS with a regular packet station using a TNC and your 2-meter radio. If you prefer, you can also just listen to 145.800 MHz in hopes of hearing one of the astronauts on voice. To call the space station, whose call is NA1SS, set your radio to transmit on 145.990 MHz and listen on 145.800 MHz. More information about Amateur Radio on the ISS is available at **www.arrl.org/ ARISS**.

satellite as directed by your tracking program. A small beam is best for starting because it will not have to be pointed very precisely. Some satellites can be contacted with simple vertical antennas when they are directly overhead and the distance to them is low. As you get better at pointing the antenna, you can use a more powerful beam and contact the satellite closer and closer to the horizon, increasing the number of stations on Earth that are in view of the satellite at the same time as your station! Always use the minimum amount of transmitter power to contact satellites, since their relay transmitter power is limited by their solar panels and batteries.

For more information on amateur satellites, investigate the Web site of AMSAT (**www.amsat.org**), the organization that coordinates the building and launch of most amateur satellites. You'll find a lot of information about how satellites work and how to find them on the air and in the sky. There are also bulletins that you can receive to update you on satellite status and news about amateur satellites. Choose one of the active satellites listed on the Web site and start your quest for a satellite QSO!

DIGITAL TECHNIQUES

Once limited to *radioteletype* (RTTY), operation using digital modes became much more popular in the 1980s when microprocessors became inexpensive enough to use in amateur equipment and the personal computer became popular. Amateurs quickly created computer and microprocessor-based equipment to send and receive not only RTTY, but other digital modes, such as AMTOR and PACTOR. The next advance in amateur digital operation has come within the past few years as the power of sound cards now found in every new computer enabled the digital modems to be moved inside the personal computer as software. This has freed amateur protocol developers to experiment and develop many new modes, such as PSK31 and MFSK. **Figure 4-9** is a reminder of how a station is set up to use the digital modes.

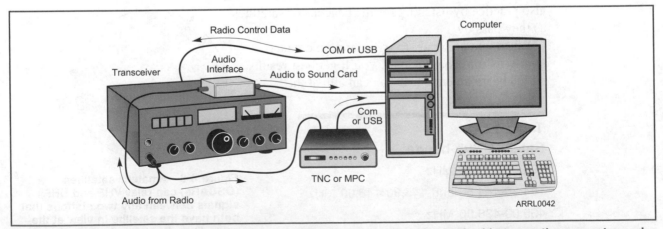

Figure 4-9—To operate on the digital modes, an external data interface is required between the computer and radio. Some external interfaces, such as packet TNCs, convert the radio's audio signals to and from data, connecting to the computer's data ports. Courtesy Wiley Publishing, Two-Way Radio and Scanners for Dummies®.

Why use digital modes, anyway? Voice and CW are quite effective but they don't have the ability of digital systems to automatically correct errors caused by noise and interference. Special codes and characters embedded in the stream of data allow the receiving modem to detect, and sometimes correct, errors. The result is that some digital modes offer error free communications at speeds that adjust automatically to propagation and noise. Protocol design is an area in which amateur experimentation is definitely advancing the state of the radio art.

Packet and Packet Networks

On VHF and UHF, the most common digital mode is *packet radio*. Most packet radio uses a protocol known as AX.25, an amateur adaptation of the wired-network X.25 protocol used for computer-to-computer communication. Data from a computer is divided into short segments and information about the data and its destination are added to form a *packet*. Packet signals are often found on simplex channels from 145.01 to 145.09 MHz. Information on packet radio can be found in the *ARRL Handbook* and the manuals for TNC data interfaces.

Packets are transmitted in bursts that sound like noise to the ear. *Frequency-shift keying* (FSK) is used to transmit the individual characters as a series of rapidly alternating audio tones. A receiving modem and Terminal Node Controller (TNC) then reassemble the data from the received packets. The characters in a packet are transmitted at 1200 or 9600 baud, so that the overall *throughput* of a packet system is about 400 or 3000 bits per second.

Individual packet stations can *connect* to each other directly, with the operators typing messages to each other, similarly to Instant Messaging. The most common type of packet use is connecting to *bulletin boards* stations. Bulletin boards accept and deliver messages, distribute bulletins, and also relay e-mail messages via the Internet. *Node* stations acting as routing centers for packet connections. Nodes are also connected to other nodes, forming networks that allow communications between individual stations through the network.

Stations can also use relays called *digipeaters* to connect to out-of-range stations. A digipeater stores the data in each packet and retransmits it. Packets can be forwarded by nodes or individual stations, as well. If the operator has a list of packet stations and their locations, the packets can be forwarded from station to station over long distances.

It is also possible to use other protocols with the same TNC equipment so that the packets are formed and handled according to a different protocol. A good example is the use of TCP/IP protocols originally developed for the Internet. TNCs programmed to use TCP/IP can form networks on the air that emulate the wired (and faster) Internet networks. This extends the Internet onto the amateur VHF/UHF bands, although at lower rates. Other amateurs are working to adapt the popular 802.11 wireless networking protocols to Amateur Radio.

Keyboard-to-Keyboard

Digital modes that are designed for live person-to-person communication are called *keyboard-to-keyboard* modes. Most popular on the HF bands, keyboard-to-keyboard mode signals are found at frequencies just above CW signals.

Radioteletype (RTTY) is the oldest, invented in the 1930s. It uses a 5-bit code for each character, called the *Baudot code* (pronounced *baw-DOH*), which is where the term *baud* comes from. RTTY originally used electromechanical marvels called teleprinters to convert the electrical current from a modem into characters. (If you ever get a chance to see one of these in action, you won't forget it!) Today, RTTY signals can be generated and received by computer software and a sound card.

There are also several "TOR" (*Teletype Over Radio*) modes used on HF; AMTOR,

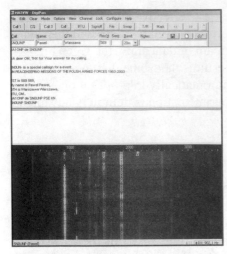

Figure 4-10—PSK31 is a popular keyboard-to-keyboard mode on HF. Many amateurs use the *DigiPan* software shown here. At the top is a window that shows the text decoded by the modem software, the bottom shows a *waterfall display* that give a visual presentation of what the computer's sound card is receiving from the radio.

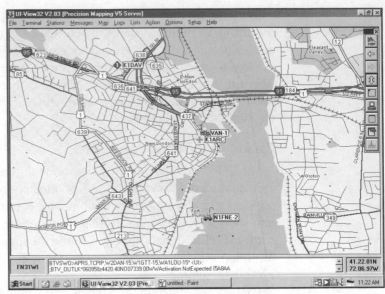

Figure 4-11—APRS, the Automatic Position Reporting System, uses packet radio signal to display the positions of fixed and mobile stations on a computer-generated map.

GTOR, PACTOR. Because of the rapid fading and signal distortion suffered by signals going through the ionosphere, data is sent in very short packets with error-correcting data to minimize the chance of error and recover from errors when they occur.

The most popular keyboard-to-keyboard mode is PSK31, which stands for *Phase Shift Keying, 31 Baud*. PSK31 is a synchronous mode that uses very precise signal timing to aid the receiving modem in recovering the signal from noise and interference. Although it is not very fast, it works very well in noisy conditions when other modes stop working. Best of all, the software to use PSK31 is free and runs on most computers. The Web site **www.digipan.com** has more information on PSK31 the popular *DigiPan* software seen in **Figure 4-10**.

APRS

The *Automatic Position Reporting System* (APRS) uses packet radio to transmit the position information from a moving or portable station to a system of APRS digipeaters. These relay points forward the position information and call sign to a system of server computers via the Internet. Once the information is stored on the servers Web sites can access the data and show the position of the station on maps in various ways. **Figure 4-11** shows an example of an APRS map.

The portable station is basically a packet radio station with a *Global Positioning System* (GPS) receiver attached to the TNC. The GPS receiver outputs a stream of position data in standard format which is interpreted by the suitably-equipped TNC and transmitted in packet form. Along with position information, some stations also transmit weather information. For more information on APRS, log on to **web.usna.navy.mil/~bruninga/aprs.html** the Web site of the system's inventor, Bob Bruninga WB4APR. The ARRL also publishes *APRS—Moving Hams on Radio and the Internet* as a reference for APRS users.

Winlink

Another Amateur Radio innovation gaining considerable popularity for emergency communications and traveling hams is the Winlink system that sends e-mail over the air waves. Winlink is a combination of a radio digital mode, usually PACTOR III on HF and packet radio on VHF/UHF, and an e-mail program, Airmail, that operates similarly to common Internet e-mail programs.

An overview diagram of the Winlink system is shown in **Figure 4-12**. A complete description of the Winlink system can be found at **www.winlink.org**. Traveling hams connect to gateway stations called *Participating Mail Box Operators* (PMBO) that are in turn connected to e-mail servers via the Internet. When the traveler connects to a PMBO, any e-mail waiting on the server can be retrieved, no matter where the traveler happens to be. It is common for hams at sea to connect to PMBOs thousands of miles away.

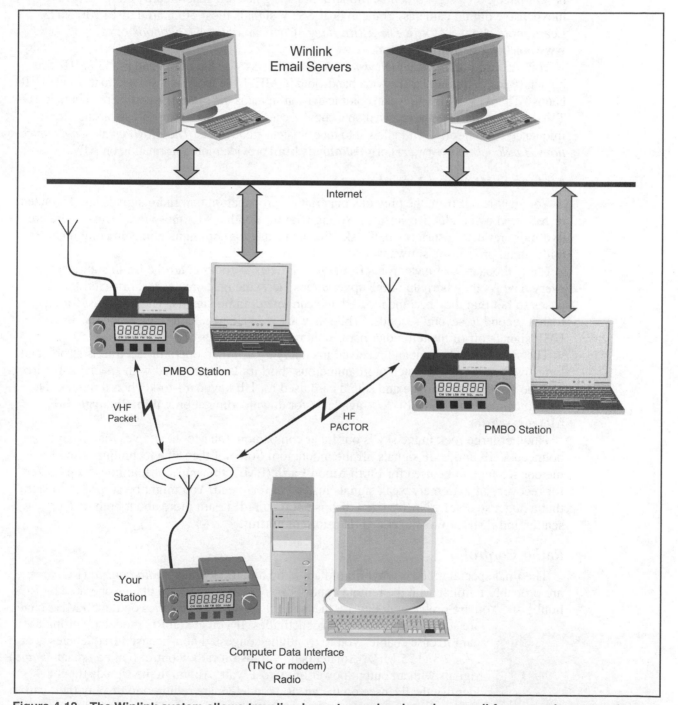

Figure 4-12—The Winlink system allows traveling hams to send and receive e-mail from anywhere over the amateur bands. PMBO stations act as gateways on HF or VHF/UHF, connecting to regular e-mail servers via the Internet. An e-mail client program, *Airmail*, provides a familiar interface on the traveler's computer.

SPECIAL MODES

Hams try almost every possible way to communicate over the radio. After all, that's a big part of what Amateur Radio is about—experimentation and adaptation. Here are a few examples of the unusual methods hams have employed.

Video

Hams have two primary means of exchanging pictures or video in real-time, aside from exchanging data files of graphic images or video. *Slow-scan television* (SSTV) was invented by hams in the 1960s in order to send still images over convention voice radios. A new image is sent once every eight seconds. Modern SSTV signals are generated by computers and inexpensive digital cameras. You can hear SSTV signals most often around 14.230 MHz. Learn more about SSTV in *The ARRL Image Communications Handbook* or at **www.arrl.org/tis/info/sstv.html**.

You can also find *Amateur Television* (ATV) enthusiasts on the UHF bands at 430 MHz and higher. (Because of the signal's wide bandwidth [6 MHz], the mode is restricted to the wide UHF bands.) The ATV NTSC fast-scan color television signal is the same as a commercial broadcast TV signal and can even be received on a commercial TV receiver equipped with a suitable frequency converter that will allow it to tune amateur frequencies. *The ARRL Image Communications Handbook* and **www.arrl.org/tis/info/atv.html** provide more information on ATV.

Meteor Scatter and Moonbounce

You may recall from the previous material on propagation that radio signals are diffracted or reflected by conductive surfaces. Along with the conductive ionosphere layers, there are two more reflecting surfaces in the sky that hams use for communications in combination with signal processing software.

Many thousands of meteoroids from dust particles to rocks enter the Earth's atmosphere every day. As they burn up in the upper atmosphere, the heat creates a short-lived trail of gases so hot that they become ionized and can reflect radio signals. The trails last for less than a second to several seconds. While they are in existence, hams can bounce VHF and UHF signals off of them to other hams hundreds or thousands of miles away.

The signals can be sent and received manually by a skilled operator if a trail is strong and long-lived enough. The much more numerous short trails can be used with special software and the same data interface and sound card used for HF keyboard-to-keyboard modes. The best-known are *HSMS* and *WSJT*, available for downloading at **en.wikipedia.org/wiki/Meteor_scatter**.

Another large rock in the sky is our lunar companion, the Moon, and yes, hams have bounced VHF and UHF signals off the moon, too! Some of the same techniques used for meteor scatter can be used for Earth-Moon-Earth (EME) transmissions, although optimized for recovery of extremely weak signals instead of for speed. You might be surprised to learn that a huge, steerable, NASA-sized dish is not required! Learn more about both meteor scatter and EME at **www.arrl.org/tis/info/moon.html**.

Radio Control

The final special mode is a hybrid of ham radio and modeling—*radio control* (RC). You are probably familiar with the remote control cars, trucks, boats, and planes operated by RC hobbyists. You are probably not aware that special amateur frequencies on the 50 MHz band are set aside for radio control activities. If you are an RC modeler, getting a ham license enables you to avoid the congested non-licensed frequencies near 27, 72 and 75 MHz. Amateurs may transmit radio control (called *telecommand*) signals with an output power of up to 1 watt. Although the signals do not identify the licensee on the air, RC modelers are required to display their call sign and address on the RC transmitter. If you would like to find out more about Amateur Radio and RC modeling, there is more information in *The ARRL Handbook* and at RC hobby shops.

Study questions
T5D13; T6A11;
T6C01-04,06,07;
T7A05-07,10-12;
T7B01-11.

Chapter 5

Licensing Regulations

In this section, you'll learn about:

- **How FCC rules are identified**
- **Ham Radio's "Mission"**
- **Types of Licenses**
- **Licensing Exams and Volunteer Examiners**
- **Responsibilities of Licensees**
- **Frequency and Emission Privileges**
- **International Radio Rules**
- **Amateur Call Signs**

It's time for the rules of the road, ham radio style! In the preceding sections, you've learned the technology and customs of Amateur Radio. Now you have the background to understand what the rules and regulations are intended to accomplish. That background will make it a lot easier for you to learn (and remember!) the rules.

There are two sections of the book that deal with the rules and regulations. This section deals with licensing regulations—bands and frequencies, call signs, international rules, how the licensing process works, etc. These are administrative rules. The next section will deal with rules about operating.

5.1 Licensing Terms

In dealing with rules and regulations, it's very important everyone uses the same words to mean the same things, so we'll begin with a series of definitions. They are, after all, what the rules are built from! In case you want to look up a specific rule or definition, they're all on line at **wireless.fcc.gov/rules.html**. Better yet, get a copy of *The ARRL's FCC Rule Book* that not only contains the latest Amateur Radio rules, but explanations and rationale so you know what is intended.

PART 97

Each of the radio services administered by the FCC has its own section of rules and regulations. The Amateur Service is defined by and operates according to the rules in Part 97 of the FCC's rules. (The FCC rules are part of Title 47 of the Code of Federal Regulations (CFR), the section on Telecommunications.) Each rule is defined in a separate section and receives a number beginning with 97, such as 97.101, a set of General Standards. Within each rule, individual parts get additional designators, for example 97.101(a) specifying good practice be used. When a specific rule is referenced, it will be enclosed in square brackets, such as [97.3 (a) (4)]. This looks complicated at first, but actually helps you find the exact rule quickly.

Basis and Purpose

Rule 97.1 is the most important rule of all—it's the *Basis and Purpose* of Amateur Radio. This explains the "mission" of Amateur Radio, why we are allocated precious RF spectrum and what Amateur Radio is intended to accomplish. Here's what Part 97.1 says, with a little explanation added:

The rules and regulations in this part are designed to provide an Amateur Radio service having a fundamental purpose as expressed in the following principles:

(a) Recognition and enhancement of the value of the amateur service to the public as a voluntary noncommercial communication service, particularly with respect to providing emergency communications.

An important word to remember is "noncommercial." Hams aren't allowed to be paid for their services (with a few exceptions), operating on a voluntary basis. That includes talking about or promoting one's business activities over the air. Along with the volunteerism inherent to the service, hams are also extremely valuable when they respond to emergencies and disasters to provide temporary communications. In fact, responding to emergencies may be the most important that the Amateur Service exists today—it is, after all the very first reason given!

(b) Continuation and extension of the amateur's proven ability to contribute to the advancement of the radio art.

Hams have a history of discovering and inventing that continues today. After World War I, hams were given all of the "worthless" short wave bands, but soon discovered that they were perfect for long-distance communications against the advice of all the experts. Even with all the communications research going on around the world, hams still invent useful systems and antennas. Ham radio's famous creativity pays back the public's investment of spectrum many times over.

> Hams have a history of discovering and inventing that continues today.

(c) Encouragement and improvement of the amateur service through rules which provide for advancing skills in both the communication and technical phases of the art.

Not only do hams tinker with radios, but they train to use them in useful ways. Events such as Field Day and the myriad exercises held all around the world are ways in which amateurs keep their emergency response skills sharp. Casual operating and station-building continually develop the ham's communications skills, too.

(d) Expansion of the existing reservoir within the Amateur Radio service of trained operators, technicians, and electronics experts.

Having a bunch of folks around that are handy with radios has turned out to be a great idea over the years! There is a long list of ways in which hams have shown their communications skills to be a valuable resource to the public, to the military, and to private industry.

(e) Continuation and extension of the amateur's unique ability to enhance international goodwill.

It has been said that ham radio is an international "Passport to Friendship." There is nothing like a live connection with someone far away, whether it be by Morse code on the HF bands or via an IRLP chat between two handheld-wielding hams. Hams are almost unique in their ability to "make contact" with people around the world every day, from all walks of life with little or no intervening systems or bureaucracy.

Throughout your time in Amateur Radio, you will encounter many different activities and events. As long as they satisfy one or more of these criteria, then, yes, they are "real" ham radio. We have by no means exhausted the possibilities!

Definitions

Let's start with the question, "Who makes and enforces the rules for the Amateur Radio Service in the United States?" This is a pretty important thing to know! The answer, of course, is the *Federal Communications Commission* or FCC. No matter what you're doing on the air, the FCC rules take precedence, even if you are operating on behalf of another government agency! The FCC is also the agency that grants your Amateur Radio license. A license in the Amateur Service allows you to operate anywhere that the FCC regulates the amateur service—the 50 states and all possessions under US government control. (You can also operate in countries with which the US has *reciprocal operating authority* as described below.)

With the FCC established as the body in charge of regulating ham radio, here are some fundamental definitions that the FCC uses to construct the rules:

Amateur Service - [97.3 (a) (4)] *"A radiocommunications service for the purpose of self-training, intercommunication and technical investigation carried out by amateurs, that is, duly authorized persons interested in radio technique solely with a personal aim and without pecuniary interest."* A telecommunications *service* is governed by a set of rules and regulations that define and administer a specific type of communications activity, such as amateur, land mobile, or marine. The Amateur Service is one of many such services.

Amateur Operator - [97.3 (a) (1)] *"A person holding written authorization to be the control operator of an amateur station."* From the point of view of the rules, "amateur" does not imply fitness or ability; it just identifies the service with whose rules the operator must comply. There are also Land Mobile Operators and Marine Operators, for example.

Amateur Station - [97.3 (a) (5)] *"A station in an amateur service consisting of the apparatus necessary for carrying on radio communications."* This sounds a bit circular, but what it means is any station that complies with the rules of the amateur service during operation.

Pecuniary - related to money or payment. This includes trade or barter! The bottom line is that you can't be paid—in money, goods, or services—for operating on behalf of another individual or an organization. (The sole exception is that clubs can pay an operator to control a station that transmits information bulletins of interest to Amateur Radio operators—such as W1AW.)

These are where it all begins; the service, the operator, and the station. There are many other definitions, of course. Some will be familiar words that may be used in unfamiliar ways. If you are in doubt about the meaning of any word in the rules, it's likely that a precise definition is already waiting for you in Part 97.3, including technical terms.

TYPE AND CLASSES OF LICENSES

Aside from passing the exam, are there any other qualifications a person must have to get an Amateur Radio license? Only one: they can't be a representative of a foreign government. (Citizens of other countries can and do get US amateur licenses.) There are no age, health, or fitness requirements—ham radio is truly equal opportunity!

An Amateur Radio license actually consists of two parts—an *operator license* and a *station license*. In most other services, they are granted separately, such as for broadcast stations where employees actually operate the equipment. The operator license gives you permission to operate an amateur station according to the rules of the Amateur Service. The station license authorizes you to have an amateur station. The combined license is an amateur *operator/primary station license*. Each person can have only one such license. **Figure 5-1** shows an actual Amateur

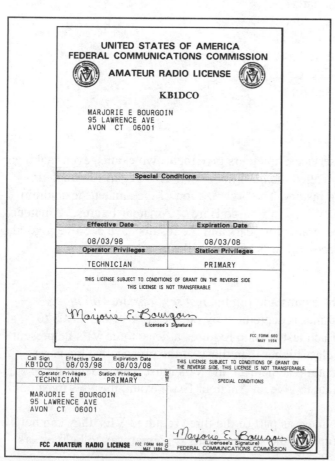

Figure 5-1—An FCC Amateur Radio license is both an operator and a station license. The printed license shown here has two sections; one for posting in your station and one to carry with you.

Table 5-1
Amateur License Class Examinations

License Class	Elements Required	Number of Questions
Technician	2 (Written)	35 (passing is 26 correct)
General	3 (Written)	35 (passing is 26 correct)
	1 (5 WPM Code)	10 (passing is 7 correct)
Amateur Extra	4 (Written)	50 (passing is 37 correct)

Radio license as you will receive in the mail from the FCC. Don't lose the original—you are supposed to have it available for inspection at any time.

There are three classes of Amateur Radio licenses being granted today; Technician, General, and Extra. Each carries a different set of *frequency and operating privileges* that expand from Technician to General to Extra along with the comprehensiveness of the exams. As you pass harder exams, you get more privileges. There are other license classes—the Novice and Advanced—for which licenses are no longer being granted. There are still Novice and Advanced license holders and you'll meet them on the air.

The exams themselves are referred to as *elements*. Each license class requires that a specific element and all lower-class elements be passed. For example, to obtain the General class license, you must pass elements 1, 2 and 3 in order. **Table 5-1** (the same as Table 1-3) shows the elements that must be passed for each license class.

> **Before you go on, study test questions T1A01-04,07-09; T1C02; T2C02; and T1D02-04. Review this section if you have difficulty.**

EXAMINATIONS

Unique among the various radio services, amateurs give their own exams, even making up the questions used on the exams (called the *question pool*)! This is part of Amateur Radio's self-policing, self-regulating history. The organization of the amateur test administration is discussed in the sidebar, "Who's In Charge Here? - Amateur Exams." Amateur volunteers have run their own test programs for years and the results have been very good, with thousands and thousands of successful tests, as you will soon see.

Volunteer Examiners

The amateurs that actually give the exams and run the *test sessions* are *Volunteer Examiners* (VEs). A Volunteer Examiner is accredited by one or more of the VECs to administer amateur license exams. Each test session requires at least three VEs be present that hold an amateur license with a class higher than those of the prospective licensees. For example, to give a Technician exam (Element 2), the VEs must hold a General class license or higher. The exception is that to give an Amateur Extra exam, the VEs must hold at least Amateur Extra licenses.

Although Technicians can't be counted as part of the three required VEs, they can help process paperwork and assist with running the session. This is a great way to help others get their license.

Taking the Exam

The first step is to find a test session by contacting the VECs in your area or checking

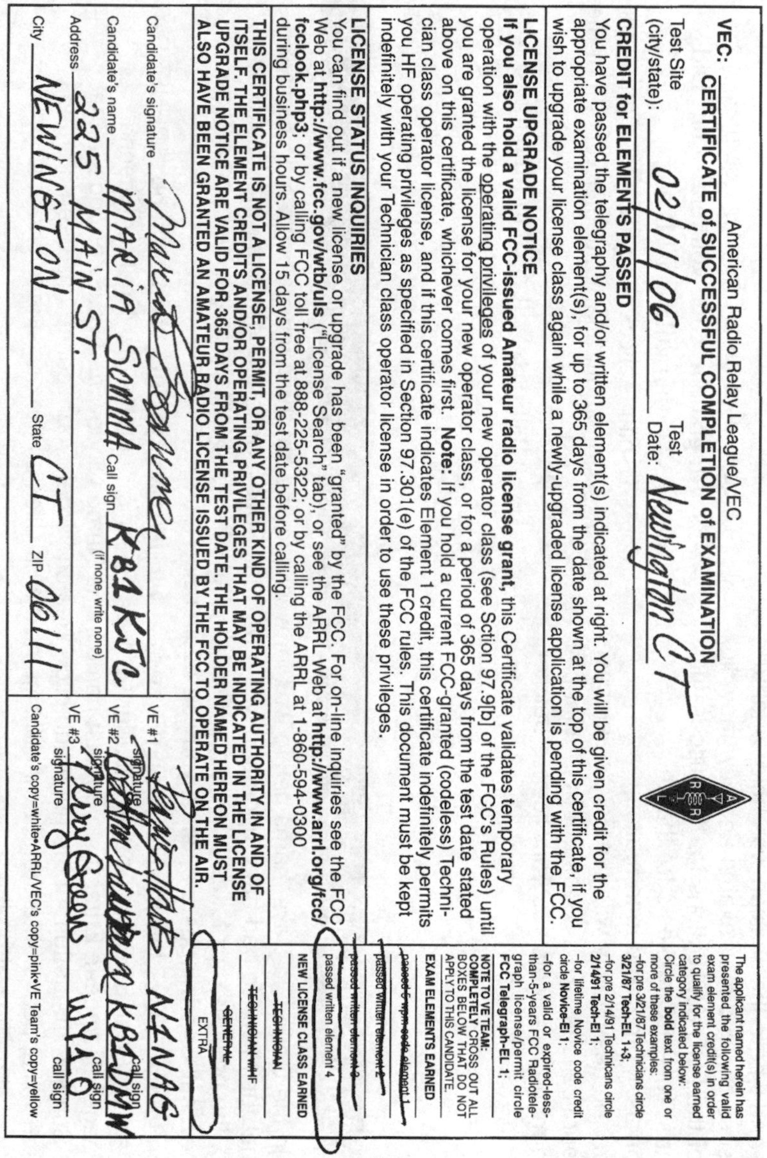

VEC: American Radio Relay League/VEC

CERTIFICATE of SUCCESSFUL COMPLETION of EXAMINATION

Test Site (city/state): 02/11/06 Test Date: Newington CT

CREDIT for ELEMENTS PASSED

You have passed the telegraphy and/or written element(s) indicated at right. You will be given credit for the appropriate examination element(s), for up to 365 days from the date shown at the top of this certificate, if you wish to upgrade your license class again while a newly-upgraded license application is pending with the FCC.

LICENSE UPGRADE NOTICE

If you also hold a valid FCC-issued Amateur radio license grant, this Certificate validates temporary operation with the operating privileges of your new operator class (see Sction 97.9[b] of the FCC's Rules) until you are granted the license for your new operator class, or for a period of 365 days from the test date stated above on this certificate, whichever comes first. **Note:** If you hold a current FCC-granted (codeless) Technician class operator license, and if this certificate indicates Element 1 credit, this certificate indefinitely permits you HF operating privileges as specified in Section 97.301(e) of the FCC rules. This document must be kept indefinitely with your Technician class operator license in order to use these privileges.

LICENSE STATUS INQUIRIES

You can find out if a new license or upgrade has been "granted" by the FCC. For on-line inquiries see the FCC Web at **http://www.fcc.gov/wtb/uls** ("License Search" tab), or see the ARRL Web at **http://www.arrl.org/fcc/fcclook.php3**; or by calling FCC toll free at 888-225-5322; or by calling the ARRL at 1-860-594-0300 during business hours. Allow 15 days from the test date before calling.

THIS CERTIFICATE IS NOT A LICENSE, PERMIT, OR ANY OTHER KIND OF OPERATING AUTHORITY IN AND OF ITSELF. THE ELEMENT CREDITS AND/OR OPERATING PRIVILEGES THAT MAY BE INDICATED IN THE LICENSE UPGRADE NOTICE ARE VALID FOR 365 DAYS FROM THE TEST DATE. THE HOLDER NAMED HEREON MUST ALSO HAVE BEEN GRANTED AN AMATEUR RADIO LICENSE ISSUED BY THE FCC TO OPERATE ON THE AIR.

The applicant named herein has presented the following valid exam element credit(s) in order to qualify for the license earned category indicated below: Circle the **bold** text from one or more of these examples:
- –for pre 3/21/87 Technicians circle **3/21/87 Tech-EL 1+3;**
- –for pre 2/14/91 Technicans circle **2/14/91 Tech-El 1;**
- –for lifetime Novice code credit circle **Novice-El 1;**
- –for a valid or expired-less-than-5-years FCC Radiotelegraph license/permit circle **FCC Telegraph-EL 1;**

NOTE TO VE TEAM: COMPLETELY CROSS OUT ALL BOXES BELOW THAT DO NOT APPLY TO THIS CANDIDATE.

EXAM ELEMENTS EARNED

~~passed 5 wpm code element 1~~
~~passed written element 2~~
~~passed written element 3~~
passed written element 4

NEW LICENSE CLASS EARNED

~~TECHNICIAN~~
~~TECHNICIAN w/HF~~
~~GENERAL~~
EXTRA

Candidate's signature Maria Somma
Candidate's name MARIA SOMMA Call sign KB1KJC (if none, write none)
Address 225 MAIN ST.
City NEWINGTON State CT ZIP 06111

VE #1 Perry Harts N1NAG (signature, call sign)
VE #2 Rose Ann Lawrence KB1DMW (signature, call sign)
VE #3 Perry Green WY1O (signature, call sign)

Candidate's copy=white•ARRL/VEC's copy=pink•VE Team's copy=yellow

(A)

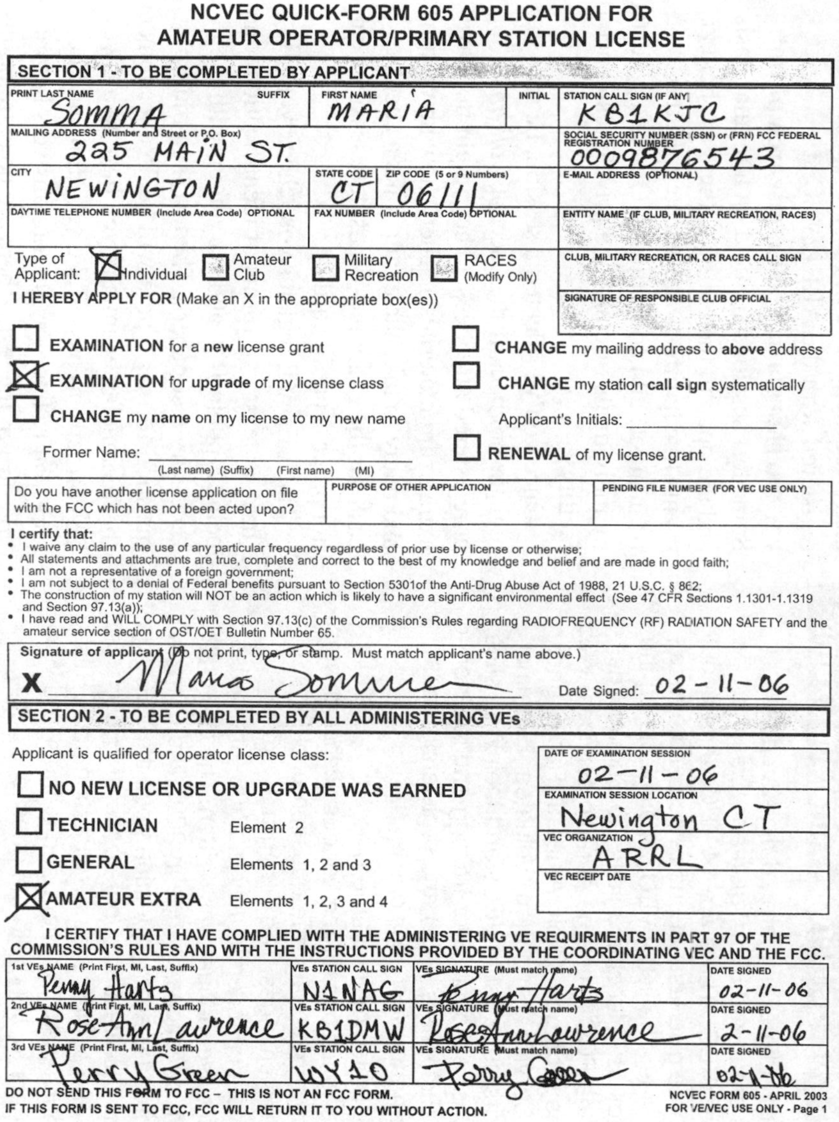

NCVEC QUICK-FORM 605 APPLICATION FOR AMATEUR OPERATOR/PRIMARY STATION LICENSE

SECTION 1 - TO BE COMPLETED BY APPLICANT

PRINT LAST NAME: SOMMA	SUFFIX	FIRST NAME: MARIA	INITIAL	STATION CALL SIGN (IF ANY): KB1KJC
MAILING ADDRESS (Number and Street or P.O. Box): 225 MAIN ST.				SOCIAL SECURITY NUMBER (SSN) or (FRN) FCC FEDERAL REGISTRATION NUMBER: 0009876543
CITY: NEWINGTON	STATE CODE: CT	ZIP CODE (5 or 9 Numbers): 06111		E-MAIL ADDRESS (OPTIONAL)
DAYTIME TELEPHONE NUMBER (Include Area Code) OPTIONAL	FAX NUMBER (Include Area Code) OPTIONAL			ENTITY NAME (IF CLUB, MILITARY RECREATION, RACES)
				CLUB, MILITARY RECREATION, OR RACES CALL SIGN
				SIGNATURE OF RESPONSIBLE CLUB OFFICIAL

Type of Applicant: [X] Individual [ ] Amateur Club [ ] Military Recreation [ ] RACES (Modify Only)

I HEREBY APPLY FOR (Make an X in the appropriate box(es))

- [ ] **EXAMINATION** for a **new** license grant
- [X] **EXAMINATION** for **upgrade** of my license class
- [ ] **CHANGE** my **name** on my license to my new name
  Former Name: ____ (Last name) (Suffix) (First name) (MI)
- [ ] **CHANGE** my mailing address to **above** address
- [ ] **CHANGE** my station **call sign** systematically
  Applicant's Initials: ____
- [ ] **RENEWAL** of my license grant.

Do you have another license application on file with the FCC which has not been acted upon?	PURPOSE OF OTHER APPLICATION	PENDING FILE NUMBER (FOR VEC USE ONLY)

I certify that:
- I waive any claim to the use of any particular frequency regardless of prior use by license or otherwise;
- All statements and attachments are true, complete and correct to the best of my knowledge and belief and are made in good faith;
- I am not a representative of a foreign government;
- I am not subject to a denial of Federal benefits pursuant to Section 5301of the Anti-Drug Abuse Act of 1988, 21 U.S.C. § 862;
- The construction of my station will NOT be an action which is likely to have a significant environmental effect (See 47 CFR Sections 1.1301-1.1319 and Section 97.13(a));
- I have read and WILL COMPLY with Section 97.13(c) of the Commission's Rules regarding RADIOFREQUENCY (RF) RADIATION SAFETY and the amateur service section of OST/OET Bulletin Number 65.

Signature of applicant (Do not print, type, or stamp. Must match applicant's name above.)
X Maria Somma Date Signed: 02-11-06

SECTION 2 - TO BE COMPLETED BY ALL ADMINISTERING VEs

Applicant is qualified for operator license class:
- [ ] **NO NEW LICENSE OR UPGRADE WAS EARNED**
- [ ] **TECHNICIAN** Element 2
- [ ] **GENERAL** Elements 1, 2 and 3
- [X] **AMATEUR EXTRA** Elements 1, 2, 3 and 4

DATE OF EXAMINATION SESSION: 02-11-06
EXAMINATION SESSION LOCATION: Newington CT
VEC ORGANIZATION: ARRL
VEC RECEIPT DATE

I CERTIFY THAT I HAVE COMPLIED WITH THE ADMINISTERING VE REQUIRMENTS IN PART 97 OF THE COMMISSION'S RULES AND WITH THE INSTRUCTIONS PROVIDED BY THE COORDINATING VEC AND THE FCC.

VEs NAME (Print First, MI, Last, Suffix)	VEs STATION CALL SIGN	VEs SIGNATURE (Must match name)	DATE SIGNED
1st: Perry Harts	N1NAG	Perry Harts	02-11-06
2nd: Rose Ann Lawrence	KB1DMW	Rose Ann Lawrence	2-11-06
3rd: Perry Green	WY1O	Perry Green	02-11-06

DO NOT SEND THIS FORM TO FCC – THIS IS NOT AN FCC FORM.
IF THIS FORM IS SENT TO FCC, FCC WILL RETURN IT TO YOU WITHOUT ACTION.

NCVEC FORM 605 - APRIL 2003
FOR VE/VEC USE ONLY - Page 1

(B)

Figure 5-2 — The CSCE (Certificate of Successful Completion of Examination) is your test session receipt that serves as proof that you have completed one or more exam elements. It can be used at other test sessions for 365 days.

Who's In Charge Here? - Amateur Exams

Amateur Radio exams are administered by volunteers organized by a *Volunteer Examiner Coordinator* (VEC). There are 14 VECs throughout the US. Some give tests in one region and others are nationwide. The list of VECs is available on the FCC's Wireless Telecommunications Bureau (the branch of the FCC in charge of radio services) Web site **wireless.fcc.gov**. Click "Amateur Radio" under "Wireless Services" and then "Volunteer Examiner Coordinators" for a list of VECs and their Web sites. Each VEC is responsible for training and registering *Volunteer Examiners* (VEs) and registering and administering the *test sessions*.

The VECs also maintain the *question pools* that contain the questions for the various license exams. Once every three years, the question pool for a license class changes. The Technician exam questions last changed in July of 2006—the reason this new license manual was written. The General class exam questions change in 2007 and Extra in 2008. To prepare a new question pool, the VECs get together as the *National Conference of VECs* (NCVEC) and publish the new question pool. This allows license class teachers and authors and Web site designers to work with the questions and prepare teaching and exam aids.

VECs also process all of the paperwork associated with the test sessions. Records are kept of when and where the session was held, what VEs ran the session, who took exams and how they did, and all of the FCC applications for the successful candidates. The VEC makes sure that the paperwork is in order and then makes the necessary entries directly with the FCC's database of amateur licensees. This saves the FCC a lot of time and money and gets you your license quicker!

their Web sites. You can find a test sessions registered with the ARRL/VEC at **www.arrl.org/arrl/vec/examsearch.phtml**. Some test sessions are held on a regular basis and are open to the public. Others are held at events like hamfests (an Amateur Radio fleamarket) and conventions. Others are hosted by individuals or clubs at a private residence or facility.

Once you've selected a session, be sure to register or check if they accept *walk-ins*, unregistered attendees. (They might not have extra exam materials.) Then show up on time and ready to take the exam. Take a check or cash for the exam fee—most test sessions can not process credit cards.

The VEs will register you, check your ID (have at least two forms of identification, one being a photo ID), and set you up with the necessary test papers. The test is in multiple-question format. There will be 35 questions on the Technician exam and you have to answer 26 correctly to pass. Take your time and be sure you're happy with your answers before you turn in the test—there are no extra points given for speed!

CSCE and Form 605

Once you've turned in your test, the VEs will grade it while you relax a bit! Did you pass? Congratulations! If not, don't despair—check with the VEs running the test session about taking another version of the exam for that element.

Those that pass will need to fill out two

Finding Your Call Sign!

Back in the Good Old Days, hams had to wait for weeks to find out their new call sign or to receive notice that their license had been upgraded to a new class. Today, once the forms are processed by the VEC, the process only takes a few days. You can check the FCC database yourself and as soon as the new call or privileges are listed, you can begin using them!

Log on to the FCC's Universal Licensing System at **wireless.fcc.gov/uls**, then click the LICENSES button next to "Search." Click "Amateur" under "Service Specific Search." The Amateur License Search page will appear. Enter just your last name in the Name window and your ZIP code in that window. Scroll to the bottom of the page and click "Search." When your license has been granted, your name will appear with a brand-new call sign next to it! Click on the call sign to check all of your information. Then go get on the air!

Obtaining Form 605

If you decide to "do the paperwork" on real paper instead of on-line, you'll need to get a blank FCC Form 605. This is not difficult! You can get FCC Form 605 with detailed instructions by contacting the FCC in any of these ways:

- FCC Forms Distribution Center, 9300 E. Hampton Dr, Capital Heights, MD 20743; tel 800-418-3676 (If you make a written request, write "Form 605" on the envelope.)
- FCC Forms "Fax on Demand" – tel 202-418-0177, ask for form number 000605
- FCC Forms On-Line - **www.fcc.gov/ formpage.html** or **ftp.fcc.gov/pub/Forms/ Form605**

The ARRL also offers a FCC Form 605 package geared to amateur needs. Write to: ARRL/VEC, FCC Form 605, 225 Main St, Newington CT 06111-1494. Include a large business-sized stamped, self-addressed envelope with your request.

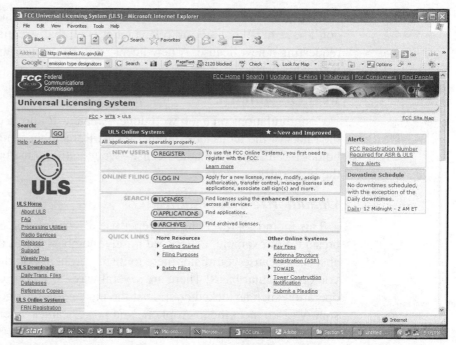

Figure 5-3—The FCC's Universal Licensing System (ULS) is the preferred method for hams to interact with the FCC. After registering, you can use the ULS to renew your license, change your address, or even change your call sign!

forms provided by the VEs; a *Certificate of Successful Completion of Examination* (CSCE) and a *NCVEC Quick Form 605* (**Fig 5-2 A-B**). The CSCE is your "receipt" of successfully passing the test for an element and what class of license you have earned. You should keep the CSCE until the FCC data base has been updated with all of your new information. The CSCE is good for 365 days as proof that you have successfully passed one or more elements.

Then you fill out a NCVEC Quick Form 605 as shown in Figure 5-2 and that will be filed with the FCC. It records your personal information and will be used to link your identity with a specific call sign. The test session VEs will help you fill out the form correctly. With both forms signed and your exam fee paid, are you a ham? Not quite—you have to wait until you are notified of your call sign as discussed in the sidebar. Once your information shows up in the FCC database, you are fully qualified to go on the air!

TERM OF LICENSE & RENEWAL

Amateur licenses are good for a 10-year term. You can renew them indefinitely without ever taking another exam. You can renew on-line by using the FCC's Universal Licensing System (ULS) at **wireless.fcc.gov/uls**. Up until 90 days before your license expires, you can also fill out a paper FCC Form 605 and mail it to the FCC. (After that time, the paperwork may not be processed in time to prevent expiration.)

What if your license does expire? People do forget! If your license expires, you are supposed to stop transmitting because your license is not valid after it expires. Nevertheless, you have a 2-year grace period to apply for a new license without taking the exam again.

If your license is lost or destroyed, you can request a new one from the FCC. You don't have to fill out a Form 605, a letter is all that's required explaining why you are requesting a new license. A new license will be printed and mailed to the address on file with your license.

RESPONSIBILITIES

Congratulations on a job well done, Amateur Radio licensee! You are ready to join the ranks of the thousands of other hams! Remember your primary responsibility as the holder of an Amateur Radio license—your station must be operated in accordance with the FCC rules.

Unauthorized Operation

This includes preventing improper use of your station equipment when you're not present. For example, unlicensed family members aren't allowed to operate in your absence because you're not there to insure proper operating—even if they operate with you on a regular basis! There are several ways to secure your station from unauthorized operation such as locked doors, but the most common and recommended methods is to simply disconnect the microphone and power cables when you're not around. That removes the temptation and prevents operation by unlicensed persons that have access to your station.

Personal Info

The FCC requires you to maintain a valid current mailing address in their database at all times. This is so that they can contact you by mail, if needed. If you move or even change PO boxes, be sure to update your information using the FCC ULS online system. If you do not maintain a current address and mail to you is returned to the FCC as undeliverable, your license can be revoked and removed from the data base.

The other piece of information that might be unfamiliar is the FRN (Federal Registration Number). This is a number assigned to you as a licensee. You can use your Social Security Number as the Taxpayer ID Number. Registering with the FCC is covered in the next chapter.

Station Inspection

As a federal licensee, you are obligated to make your station available for inspection upon request by an FCC representative or US government official. By accepting the FCC rules and regulations for the Amateur Service, you agree that your station could be inspected any time. These visits are very rare and only occur when there is reason to believe that your station has been operated improperly. Remember to keep your original license available for inspection, too!

Before you go on, study test questions T1A04-06; T1D05-12; T2D08-10.

5.2 Working with the FCC

While you can still use paper forms and letters to interact with the FCC, working on-line is much quicker and often simpler. As you become accustomed to working on-line, you can also tap a wealth of information stored in the FCC database.

THE FCC ULS WEB SITE

The FCC has developed a comprehensive Web site for all of its licensees to use: It's called the *Universal Licensing System* or ULS and the Web address for it is **wireless.fcc.gov/uls**. The ULS home page is shown in Figure 5-3. You can use the ULS in several ways:

- Register for on-line access to your license information
- Make simple changes to your address and other information
- Renew your license
- Search for licensees by name, call sign, or location

If you become responsible for more than one call sign, such as for a club or repeater, the on-line system is much easier to deal with.

Here's a sample of how to use the ULS Search function. Browse to the site and click the LICENSES button next to "Search." When the License Search page loads, click on "Service Specific Search". Click "Amateur" and when the search form loads, enter your ZIP code and click "Search" at the bottom of the page. You might be surprised at the results!

REGISTERING

To use the ULS site for managing your license, you'll need to register with the FCC. You can register whether or not you have a license. Registering is done through *CORES*—the Commission Registration System. Click the REGISTER button next to "New Users" on the ULS home page to begin the registration process. (The ARRL has a complete registration guide on-line at **www.arrl.org/fcc/uls101.html**.) You will be assigned a unique FCC identification number, your *Federal Registration Number* or FRN.

> The FRN is your key to unlocking all of the license management services available via the ULS system.

USING YOUR FRN

The FRN is your key to unlocking all of the license management services available via the ULS system. You must use it on paper forms, such as FCC Form 605. Having an FRN means you don't have to use your Social Security Number when filing forms. When you register, you will receive a letter that contains your FRN and a starting password. Log on to the ULS site and click the LOG IN button next to "Online Filing." Enter the FRN and password in the letter. Once logged in you can then change your password to something you prefer (and can remember!). Don't lose either your FRN or your password!

If you become the FCC's contact for club or repeater call signs, you must set up a unique FRN for the club or repeater. Do not use your individual FRN number.

5.3 Bands and Privileges

The frequencies and modes and methods that hams are allowed to use are all known to the rules and regulations as *privileges*. What gives the FCC authority to grant privileges is the Communications Act of 1934. What then grants these privileges to you is your license. By signing Form 605 and applying for a license, you agree to be bound by the FCC rules and that means staying within the privileges of your license.

To stay within those privileges, you have to know what they are! You will have to memorize some of the frequency limits for the different bands. Relax—the exam expects you to remember the most common bands and not every one. Hams keep a chart of their privileges handy, since not many of us have them all memorized. To help you remember your privileges, copy the information onto a piece of paper and tape it in your car or near your computer or on the refrigerator. Take every opportunity to recite the information, reinforcing it time after time. You'll find that it's not nearly as hard as it seems as first!

FREQUENCY PRIVILEGES

The most important privileges are *frequency privileges*. Once upon a time, at the dawn of radio, hams and broadcasters and commercial and military stations were all mixed in together in one big band. That didn't work very well when the one big band got crowded. Soon, the powers that be decided to shoo the hams off to the "worthless" frequencies with wavelengths shorter than 200 meters (1.5 MHz and higher). That worked fine for hams because they soon discovered that those frequencies were precisely the ones that supported the best long-distance communications! That couldn't last long and it didn't; soon the short wave bands were carved up among the different users and hams had their first "bands".

Today, there are literally hundreds of bands and dozens of different types of radio spectrum users. The frequency privileges granted to the various services are called *allocations*. For example, amateurs are allocated 144-148 MHz, the 2-meter band. **Figure 5-4** is a grand overview of the radio spectrum and where amateurs have frequency privileges. (If you want to get a look at the full chart of allocations, browse to **www.ntia.doc.gov/sosmhome/allochrt.html** - it's quite an eyeful!) From Figure 5-4, you can see that amateur allocations are sprinkled throughout the radio spectrum, not concentrated in or excluded from any one region. As you recall from our discussion on propagation, radio signals of different frequencies propagate differently. Thus, it's fortunate that spectrum planning has resulted in amateurs having access to a wide range of frequencies in which to both experiment and apply to different communications needs.

Technician class frequency privileges are all above 50 MHz. **Table 5-2** shows the Technician frequency privileges that you are expected to know for your license exam. Remember that a band can be referred to by frequency ("50 MHz") or by wavelength ("6 meters"). Use the formula f (in MHz) = 300 / wavelength (in meters) or wavelength (in meters) = 300 / f (in MHz) to convert between frequency and wavelength.

Not all frequency privileges are granted on a worldwide basis. The FCC doesn't have authority outside the US—every country has its equivalent

Table 5-2
VHF and UHF Technician Amateur Bands
ITU Region 2

Band (Wavelength)	Frequency Limits
VHF Range	
6 meters	50 - 54 MHz
2 meters	144 - 148 MHz
1.25 meters	219 - 220 MHz
1.25 meters	222 - 225 MHz
UHF Range	
70 centimeters	420 - 450 MHz
33 centimeters	902 - 928 MHz
23 centimeters	1240 - 1300 MHz
13 centimeters	2300 - 2310 MHz
13 centimeters	2390 - 2450 MHz

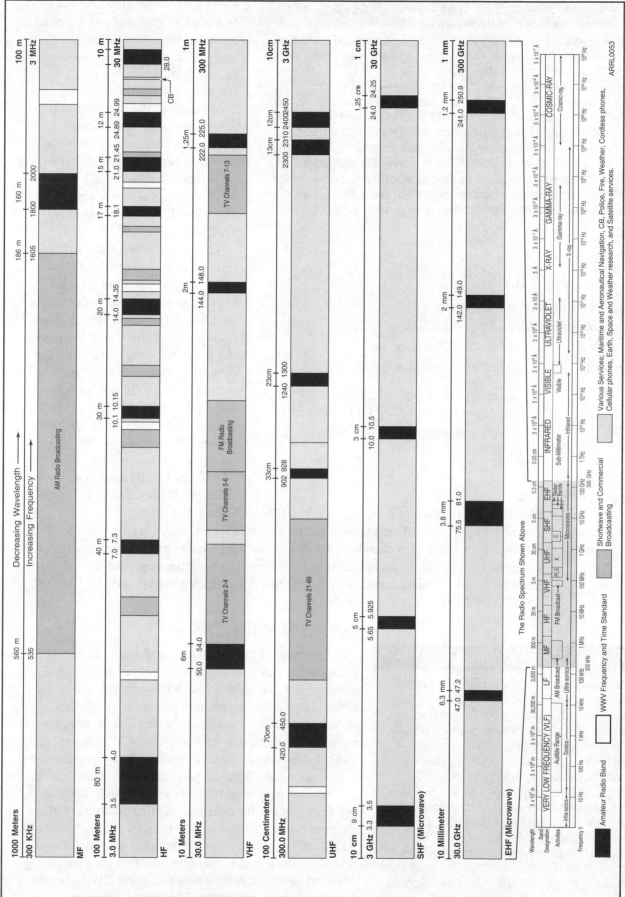

Figure 5-4 — Amateur allocations, the black rectangles in the chart, are distributed relatively evenly throughout the radio spectrum. The variations in propagation at these different frequencies give hams a lot of opportunities to experiment with different types of communications systems and methods.

agency. How do they coordinate? It would be chaos if every country made up their own allocations, since radio waves don't stop at international borders! Realizing the need for international coordination, the *International Telecommunications Union* (ITU) was formed to provide a forum in which countries could work out allocations and many other issues. (You'll learn more about the ITU later on in this section.)

The ITU divides the world into three regions—North and South America form Region 2—and organizes allocations accordingly. Because VHF and UHF signals do not frequently travel beyond the radio horizon, it was decided that allocations in these frequency ranges did not have to align precisely. As a result, the amateur allocations vary between regions. For example, 50-54 MHz is allocated to TV channel in Europe instead of Amateur Radio. Table 5-2 shows allocations for Amateur Radio above 50 MHz in Region 2.

Within the amateur HF bands, there is an additional subdivision by license class. Starting with the Novice, as higher class licenses are obtained more and more frequency privileges are granted until as an Amateur Extra class licensee all amateur privileges are granted. For example, on the 80-meter band, Novices (and Tech Plus license holders) may use CW from 3.525-3.600 MHz.

EMISSION PRIVILEGES

Within most of the ham bands, additional restrictions are made by mode or *emission type*. (Emission is the formal name for any radio signal from a transmitter.) Just as a frequency privilege is permission to use a specific frequency, an *emission privilege* is permission to communicate using a particular mode, such as phone, CW, data, or image. **Table 5-4** lists the modes that can be used by amateurs—as a Technician class licensee, you can use all of them.

The combination of frequency, class, and emission privileges makes for a fairly complicated division of the amateur bands into *sub-bands*. Parts of the ham bands in which only certain modes can be used are called *mode-restricted*. As a Technician licensee, though, your situation is very simple: There is a small CW-only sub-band occupying the bottom 100 kHz of the 6- and 2-

Table 5-4

Amateur Emission Types

Emission	Description
CW	Morse code telegraphy
Data	Computer-to-computer communication modes, usually called *digital modes*
Image	Television (fast-scan and slow-scan) and facsimile or fax
MCW	Tone-modulated CW, Morse code generated by keying an audio tone
Phone	Speech or voice communications
Pulse	Communications using a sequence of pulses whose characteristics are modulated in order to carry information
RTTY	Narrow-band, direct-printing telegraphy received by automatic equipment, such as a computer or teleprinter
SS	Spread-spectrum communications in which the signal is spread out over a wide band of frequencies
Test	Transmissions containing no information

meter bands. The segment of the 1.25-meter band from 219-220 MHz is restricted to digital message forwarding only. For all amateur allocations above 220 MHz, there are no other sub-bands! **Table 5-5** shows all of the subdivisions of amateur bands through 23 cm.

Why have mode-restricted sub-bands? Because the methods of operating for the different modes are sometimes not compatible. CW and phone operation, for example, are conducted quite differently and the signals interfere with each other. In the past few years, with the increasing number of digital modes, incompatibilities between digital and CW signals are causing interference between these two groups of operators. Hams have worked around this problem voluntarily by using narrow-bandwidth modes, such as CW, at the low-frequency end of the bands and wider-bandwidth signals, from data through voice, higher in the band. As more and more signals, even voice, appear on the air in digital form, the FCC is considering proposals to change the subdivision of bands by mode to subdividing by signal bandwidth. (Check the ARRL Web site for more information after this book is published.) A formal division of the bands by bandwidth would help address these problems, although not eliminate them entirely. That is the price of flexibility to experiment and use all the different modes!

POWER LIMITS

The maximum power an amateur is allowed to generate at the output of the transmitter or amplifier is 1500 Watts of *Peak Envelope Power* (PEP). PEP is the maximum power during one RF cycle of the radio signal at the very peak of a modulating waveform, such as for speech. For a CW signal, PEP is measured during the *key-down period* in which the transmitter is ON. FM, you will recall, is a constant-power mode, so it does not matter whether you are speaking or not. Transmitter output power is measured at the output of the last amplifier, whether internal to the transmitter or an external piece of equipment, at the input to the antenna feed line—not at the antenna or anywhere along the feed line. Fifteen-hundred watts is allowed nearly everywhere on the ham bands except on the following frequencies:

- In the Novice sub-bands on 80-, 40-, and 15-meters all amateurs are limited to 200 watts PEP

Table 5-5

The amateur bands are divided by license class and mode. Segregation of the different modes helps prevent interference between incompatible types of operating and signals. The increase in frequency privileges with class is an incentive for hams to upgrade their skills and knowledge of radio.

US Amateur Bands

ARRL *The national association for* **AMATEUR RADIO**

160 METERS

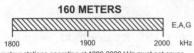

E,A,G

1800   1900   2000   kHz

Amateur stations operating at 1900-2000 kHz must not cause harmful interference to the radiolocation service and are afforded no protection from radiolocation operations.

80 METERS

3525 3600   3800

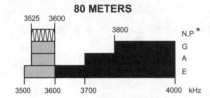

N,P *
G
A
E

3500   3600   3700   4000 kHz

60 METERS

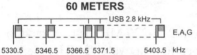

USB 2.8 kHz

E,A,G

5330.5 5346.5 5366.5 5371.5   5403.5 kHz

General, Advanced, and Amateur Extra licensees may use the following five channels on a secondary basis with a maximum effective radiated power of 50 W PEP relative to a half wave dipole. Only upper sideband suppressed carrier voice transmissions may be used. The frequencies are 5330.5, 5346.5, 5366.5, 5371.5 and 5403.5 kHz. The occupied bandwidth is limited to 2.8 kHz centered on 5332, 5348, 5368, 5373, and 5405 kHz respectively.

40 METERS

7025   7125

7175

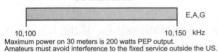

N,P *
G †
A †
E †

7000   7125   7300 kHz

† Phone and Image modes are permitted between 7075 and 7100 kHz for FCC licensed stations in ITU Regions 1 and 3 and by FCC licensed stations in ITU Region 2 West of 130 degrees West longitude or South of 20 degrees North latitude. See Sections 97.305(c) and 97.307(f)(11). Novice and Technician Plus licensees outside ITU Region 2 may use CW only between 7025 and 7075 kHz. See Section 97.301(e). These exemptions do not apply to stations in the continental US.

30 METERS

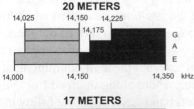

E,A,G

10,100   10,150 kHz

Maximum power on 30 meters is 200 watts PEP output. Amateurs must avoid interference to the fixed service outside the US.

20 METERS

14,025   14,150   14,225

14,175

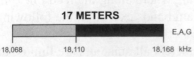

G
A
E

14,000   14,150   14,350 kHz

17 METERS

E,A,G

18,068   18,110   18,168 kHz

15 METERS

21,025   21,200

21,275

21,225

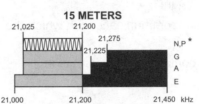

N,P *
G
A
E

21,000   21,200   21,450 kHz

12 METERS

E,A,G

24,890   24,930   24,990 kHz

10 METERS

28,000   28,500

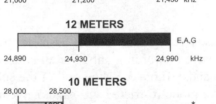

N,P *
E,A,G

28,000 28,300   29,700 kHz

6 METERS

50.1

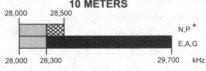

E,A,G,P,T *

50.0   54.0 MHz

2 METERS

144.1

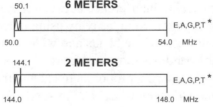

E,A,G,P,T *

144.0   148.0 MHz

1.25 METERS ***

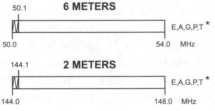

E,A,G,P,T,N *

222.0   225.0 MHz

Novices are limited to 25 watts PEP output from 222 to 225 MHz.

70 CENTIMETERS **

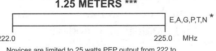

E,A,G,P,T *

420.0   450.0 MHz

33 CENTIMETERS **

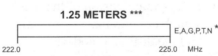

E,A,G,P,T *

902.0   928.0 MHz

23 CENTIMETERS **

1270   1295

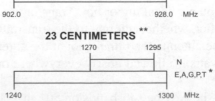

N
E,A,G,P,T *

1240   1300 MHz

Novices are limited to 5 watts PEP output from 1270 to 1295 MHz.

US AMATEUR POWER LIMITS

At all times, transmitter power should be kept down to that necessary to carry out the desired communications.

Power is rated in watts PEP output. Unless otherwise stated, the maximum power output is 1500 W.

Power for all license classes is limited to 200 W in the 10,100-10,150 kHz band.

Novices and Technicians are restricted to 200 W below 28.5 MHz.

In addition, Novices are restricted to 25 W in the 222-225 MHz band and 5 W in the 1270-1295 MHz subband.

KEY

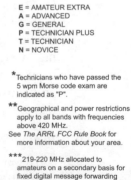

= CW, RTTY and data

= CW, RTTY, data, MCW, test, phone and image

= CW, phone and image

= CW and SSB phone

= CW, RTTY, data, phone, and image

= CW only

= USB Phone only

E = AMATEUR EXTRA
A = ADVANCED
G = GENERAL
P = TECHNICIAN PLUS
T = TECHNICIAN
N = NOVICE

* Technicians who have passed the 5 wpm Morse code exam are indicated as "P".

** Geographical and power restrictions apply to all bands with frequencies above 420 MHz.
See *The ARRL FCC Rule Book* for more information about your area.

*** 219-220 MHz allocated to amateurs on a secondary basis for fixed digital message forwarding systems only and can be operated by all licensees except Novices.

All licensees except Novices are authorized all modes on the following frequencies:
2300-2310 MHz
2390-2450 MHz
3300-3500 MHz
5650-5925 MHz
10.0-10.5 GHz
24.0-24.25 GHz
47.0-47.2 GHz
76.0-81.0 GHz
122.225-123.0 GHz
134-141 GHz
241-250 GHz
All above 275 GHz

- Technician Plus licensees are limited to 200 watts PEP in their 10-meter allocation between 28.1 and 28.5 MHz.
- All amateurs are limited to 200 watts PEP on the 30-meter band.
- All amateurs are limited to 50 watts PEP in the 219-220 MHz segment of the 1.25-meter band
- Stations being operated as beacons are limited to 100 watts PEP
- Stations operating in the 70-cm band near certain military installations may be limited to 50 watts PEP.

Most amateurs rarely use or *run* more than a few hundred watts on the VHF and UHF bands. Exceptions would be while pursuing very weak-signal methods, such as Earth-Moon-Earth (EME) or tropospheric propagation where high power is required to establish and maintain contact. High power levels at these frequencies can create safety hazards. We discuss RF safety in the final section of this book.

PRIMARY & SECONDARY ALLOCATIONS

It would be nice if every type of radio user had exclusive access to their allocations. Many amateur bands are exclusively allocated to hams, worldwide. Because spectrum is scarce and many services have valid needs for radio communications, occasionally two services receive *shared allocations,* including some of the amateur bands. When this happens, one group is generally given priority and these are called *primary allocations.* Groups that have access to spectrum on a lower priority receive *secondary allocations.* The groups that receive the allocations are *primary and secondary services.*

The primary service is *protected* from harmful interference by signals from secondary services. The secondary service gains access to the frequencies in the allocation with the understanding that it must not cause harmful interference to primary service users and it must accept interference from primary users. By sharing the bands in this way, more frequencies are available for more users than if every frequency was exclusively allocated to one service alone. Hams share several of our bands and enjoy wider access to frequencies than would otherwise be the case.

All of the UHF and higher-frequency bands have some kind of sharing arrangements. They may apply to certain geographic areas or around certain military installations. A good example is the restriction on 70-cm band operations for amateurs north of "Line A"—approximately 50 miles south of the Canadian border. Because Canada allocated 420-430 MHz to other services, to prevent international interference the FCC has ruled that US amateurs may not use that segment of the band. Frequency sharing of amateur HF bands is less common, but the 60-meter band is shared with US government services and 40-meters is shared with international short wave broadcast stations.

FCC Rule [97.303] lists all of the frequency-sharing requirements for US hams and is available on the ARRL Web site at **www.arrl.org/hrlm/sharing**. It is worth familiarizing yourself with sharing requirements to avoid interference either to or from your station!

Before you go on, study test questions T1C04-09; T3B08-10; T4B10-12; T6C05.

5.4 International Rules

In the early days of radio, every country made up its own radio rules. As equipment got better and frequencies were used that allowed long-distance communications, it became clear that wasn't going to work. After World War I, governments got together and started making international treaties that specified how the countries were to regulate radio communications. This worked reasonably well until World War II, during which communications technology made major advances. After the war was ended, a new method of managing radio was needed.

ITU (INTERNATIONAL TELECOMMUNICATIONS UNION)

With the establishment of the United Nations in 1949, the *International Telecommunications Union* (ITU) was formed as an agency of the UN. The ITU is an administrative forum for working out international telecommunications treaties and laws. The ITU also maintains international radio laws that all UN countries agree to abide by.

Regions

To manage the process of frequency allocation, the ITU has divided the world into three regions, shown in **Figure 5-5**. The US, including Alaska and Hawaii, is part of Region 2. The privilege allocation rules are fairly consistent within a single region, although there are some exceptions. For example, hams in Hawaii can operate on 40-meter phone between 7.075 and 7.100 MHz, whereas mainland hams can not.

Radio rules change at region boundaries, regardless of your citizenship or vessel ownership. If you operate from a boat, for example, when you cross from Region 2 to Region 1, you will then have to operate according to the Region 1 rules, even if you are a citizen of a Region 1 country and the boat is "flagged" in Region 3.

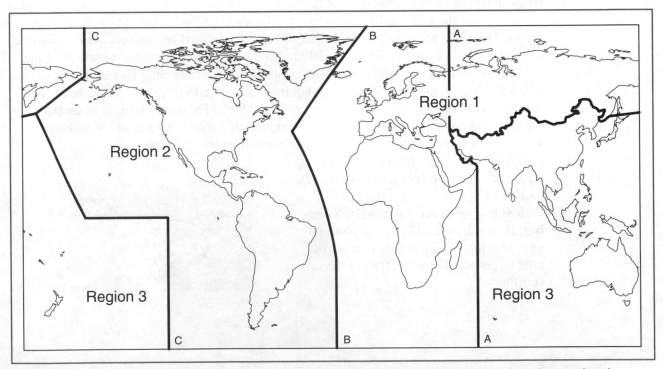

Figure 5-5 — The map shows the world divided into the three administrative regions of the International Telecommunications Union. This helps the ITU and member nations manage frequency allocations a round the world.

INTERNATIONAL OPERATING

Operating in a foreign country can be a lot of fun! You can meet local hams and if you are licensed for HF operation, become "DX" and attract a crowd on the air. To operate, you must have permission and when you are inside a country's national boundaries, including territorial waters, you are required to operate according to their rules. There are three ways of getting operating permission; reciprocal operating authority, an International Amateur Radio Permit (IARP), and the European Conference of Postal and Telecommunications Administrations (CEPT) license agreement. For more information on these agreements, information is available on the ARRL's Web site at **www.arrl.org/FandES/field/regulations/io/#us**.

Regardless of which avenue is available, don't forget the final part of the Amateur service's basis and purpose; to foster goodwill. You are a ham radio ambassador while on the air!

RECIPROCAL OPERATING AUTHORITY

Many countries have entered into *reciprocal operating authority* agreements with the United States, recognizing each other's amateur licenses. This is a government-to-government agreement recognizing each other's amateur licenses. No additional information is required; just take a copy of your US ham license with you.

IARP

In some North and South American countries, an IARP allows US amateurs to operate without seeking a special license or permit to enter and operate from that country. The IARP is issued by a member-society of the *International Amateur Radio Union* (IARU)— for the US, the IARU member society is the American Radio Relay League (ARRL). The IARP can either be Class 1 (equivalent to the US Extra) or Class 2 (equivalent to the US Technician).

CEPT

A CEPT license allows US Amateurs to travel to and operate from most European countries or their possessions without obtaining an additional licensee or permit. When traveling to a CEPT country, you'll need to have your original US license, proof of US citizenship such as a Passport, and a copy of the FCC's Public Notice about CEPT licenses. The two CEPT license classes are the same as described for the IARP.

PERMITTED CONTACTS

Unless specifically prohibited by the government of either country, any ham can talk to any other ham. Some countries do not recognize Amateur Radio, although the number is very small. Other countries prohibit contacts between their citizens and those of specific other countries. Again, this is quite uncommon.

Inside the US, the FCC may impose restrictions on a US ham as part of a judgment or administrative ruling. This is unusual and most often the result of some kind of purposeful bad behavior on the part of the limited amateur. Remember that in a communications emergency, you can still talk to any ham anywhere, if needed to prevent loss of life or property.

International Amateur Radio Union (IARU)

Just as the countries of the world support the ITU, so do the amateur "countries" support the IARU. Each country with a national society, such as the ARRL in the US, is part of the IARU, which is organized in three regions, just like the ITU. Formed in 1925 as national governments began forming radio law, the IARU (**www.iaru.org**) acts as the world-wide amateur voice to government and international rules making bodies, such as the ITU.

Before you go on, study test questions T1B01,02,07; T1C10.

5.5 Call Signs

This is the most enjoyable part of international radio rules! Call signs are our "radio names" and each amateur's is unique. Call signs all have a common structure and once you learn it, figuring out the nationality of call sign (or *call*) is easy.

PREFIX AND SUFFIX

Every call has a *prefix* and a *suffix*. The prefix consists of two or three letters and numerals. Every country is assigned at least one unique *block* of prefixes. For example, US amateur call signs begin with K, N, W, or the two-letter combination AA through AL. No matter what, if you hear a call sign beginning with those letters, you know it's a US call sign. Canadian hams use VA through VZ, French hams use F, Japanese amateur call signs begin with a J, and hams from Singapore have calls beginning with 9V.

The suffix of a call sign is the unique part that identifies the particular station and consists of only letters. In the call W1AW, "AW" is unique among all other calls beginning with "W1" (known as "W1 calls"). Suffixes are one, two, or three letters. The combination of prefix and suffix uniquely identify a station anywhere on Earth. Within a country, the call signs can be assigned to indicate license class or location or other special characteristics. For example, Russian calls beginning with the letter "U" and a single numeral are granted only to World War II veterans!

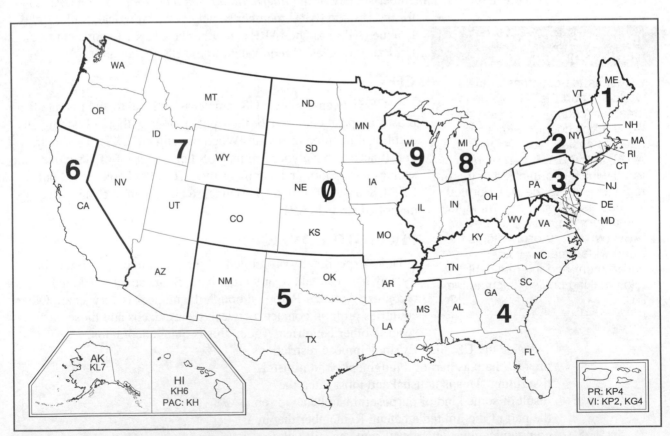

Figure 5-6 — There are ten US call districts in the continental United States. When an amateur passes a license exam or upgrade, the numeral of the district in which he or she resides becomes part of their assigned call sign. Alaska and the US possessions in the Pacific and Caribbean form three more districts and have calls with unique prefixes.

US Call Districts and Call Signs

In the US, the number in the call sign's prefix indicate in which one of the ten districts shown in **Figure 5-6** the call was originally assigned. The call sign is permanently assigned to the individual operator in the US and remains the same, no matter where the operator moves in the country. W3IZ for example, originally got his call in the third district, but now lives in Connecticut.

A US call sign is further classified by the number of letters in the prefix and suffix. A call such as WA6CF with a two-letter prefix, "WA", and a two-letter suffix, "CF", is called a *2-by-2 or 2x2*. WA1ZMS is a 2x3. W1A is a 1x1. US call signs have only one numeral in the prefix. The FCC grants these calls by license class as shown in Table 5-6. When you receive your Technician license, you will receive a Group C call sign, if any are available in your district. If all Group C calls are gone, then you will receive a Group D call. Calls are always assigned in sequential order—you get the next one on the stack!

PORTABLE & UPGRADE

Assuming your call was granted in the continental US, you can operate anywhere within the US with the same call. You are properly identified as a US ham on the air and there is no requirement for you to indicate where in the country you happen to be. However, hams often add a *portable designator* following the call sign if they are operating somewhere outside the district that would otherwise be indicated by the numeral in the prefix. For example, if NØAX operates while driving through California, he might give his call as NØAX/6, pronounced "N Zero A X portable-6" (or NØAX/6 on CW) on the air.

If you are operating outside the continental US, you must add the prefix of the country to your call because you are no longer in your home country. If you travel to Canada, for instance, you are then required to sign your call and follow it with "portable VE#" where the # indicates the Canadian province in their call sign system. If you go to Hawaii, a US state, but not part of the continental US, you would sign your call followed by "portable KH6". Similar requirements exist in other countries. Check their rules for how you are supposed to identify on the air.

Table 5-6
US Amateur Call Signs in Regions 0-9

Group	License Class	Format
Group A	Amateur Extra	Prefix K, N, or W with two-letter suffix (1x2), or two-letter prefix beginning with A, N, K, or W and one-letter suffix (2x1), or two-letter prefix beginning with A and a two-letter suffix (2x2)
Group B	Advanced	Two-letter prefix beginning with K, N, or W and a two-letter suffix (2x2)
Group C	General, Technician, and Technician Plus	One-letter prefix beginning with K, N, or W and a three-letter suffix (1x3)
Group D	Technician, Novice, and Club	Two-letter prefix beginning with K or W and a three-letter suffix (2x3)

Hams often append a designator to indicate that they are engaging in a particular activity, such as adding "/M" or saying "Mobile" after your call when operating from a moving vehicle. These *self-assigned* designators are allowed, as long as they are not the same as a designator that would conflict with the prefix of another country. For example, adding "/VE" to your call because you are a Volunteer Examiner is not OK because it would make listeners think you were in Canada!

When you upgrade your license, as soon as you receive your CSCE from the VEs administering the exam, you can append a portable designator to indicate your new license class. When your new license class appears in the FCC database, you can drop the upgrade designator. Here are the three designators:

- Technician Plus - say "portable KT" or send /KT on CW
- General - say "portable AG" or send /AG on CW
- Amateur Extra - say "portable AE" or send /AE on CW

The designator lets other stations know that you have upgraded your license although the database does not yet show the change.

CHOOSING A CALL SIGN

You can also choose your own call sign! You can have almost as much fun choosing a call from the *vanity call* program as in choosing a vanity license plate. Many hams pick a call with their initials in the suffix or that form a short word. You can pick any available call that is authorized for your license class. That means a Technician can't pick a Group A call for Extras. Nevertheless, there are lots of available calls to choose from. You can find out more about the vanity call program at **www.arrl.org/hrlm/vanity**.

CLUB AND SPECIAL EVENT CALL SIGNS

Clubs can also have their own call signs. There are some rules about the club, however. The club must have at least 4 members and the FCC can ask for documentation showing that the club exists and has meetings. Club licenses are granted to the person the club designates as the *trustee* of the club station. Your club can apply for a call sign by contacting a Club Station Call Sign Administrator. Once the club call has been assigned from Group D, the club can also use the vanity call program to change the call sign.

Any FCC-licensed amateur or club can also obtain a special 1×1 (1-by-1) call sign such as W3X or K6P for a short-duration special event. Application for the call must be made to a Special Event Call Sign Coordinator (the ARRL is one) and a call will be granted for a short time, usually 15 days. The usual call of the amateur or club requesting the special event call must be given once per hour and the special event call used for regular identification. Special event call signs are very popular and contacts with them are sought out by hams, worldwide.

> *Before you go on, study test questions T1B03-06,08-10; T2B08,09,11; T2D06.*

Chapter 6
Operating Regulations

In this section, you'll learn about:

- **Control operators**
- **Guest operating and privileges**
- **Identification on the air**
- **Tactical call signs**
- **Rules about interference**
- **Third-party communications**
- **Remote and Automatic control**
- **Prohibited communications**
- **Broadcasting**

You might be surprised to learn that there really aren't very many regulations about operating! The FCC relies on hams to come up with procedures and methods that work. As long as ham's signals meet the technical requirements and these few operating rules are followed, the FCC trusts Amateur Radio to be self-regulating. Contrary to what you might think, the FCC steps in only rarely to resolve disputes or otherwise intractable problems. As a result, hams have a great deal of freedom to innovate and adapt within the rules with just a handful of simple do's and don'ts.

6.1 Control Operators

The primary concern of the FCC is that all transmissions be made under the control of a properly-licensed operator who is responsible for making sure all FCC rules are followed. That operator is the station's *control operator*. There can be only one control operator for a station at a time. That's the person responsible for station operation, no matter who is actually speaking into the microphone.

DEFINITIONS

Reading the rules themselves can be a bit confusing because of the language used, so let's be sure to clearly define the terms. There are two basic ideas on which amateur rules are based: a control operator responsible for creating a signal and a control point at which control is asserted by the control operator.

- A control operator is the amateur designated as responsible for making sure that transmissions comply with FCC rules. The control operator doesn't have to be the station owner and doesn't even have to be physically present at the transmitter, as you will see below, but all amateur transmissions are the responsibility of a control operator.
- A control operator must be named in the FCC amateur license database or be an alien with reciprocal operating authorization. (An alien is a citizen of another country.) This is a simple requirement—the FCC has to know who you are, that you are licensed, and where you can be contacted. Any licensed amateur can be a control operator.
- The *control point* is where the station's control function is performed. Usually, the control point is at the transmitter and the control operator physically manipulates the controls of the transmitter. The control point can be remotely located; connected by phone lines, the Internet, or a radio link.

PRIVILEGES & GUEST OPERATING

As the control operator, you may operate the station in any way permitted by the privileges of your license class. It doesn't matter what the station owner's privileges are, only the privileges of the control operator. When the station owner and the control operator are the same person, responsibilities are easy to understand.

Being a guest operator is very common—you may allow another amateur to use your station or you may be the guest. Either way, it's important to understand what sets the control operator's privileges. A guest operator hosted by a higher-class licensee can operate using the host's privileges only if the host is the control operator. If the host is not the control operator, the guest is restricted to the privileges of their license.

Here's an example—you, a Technician-class licensee, are invited to spend the afternoon at the station of your Elmer, who holds an Extra-class license. While your Elmer is supervising, you can operate the station on any amateur band and mode. This is very common and is a good way to learn about the HF bands and styles of operating not used on VHF/UHF. However, if your Elmer decides to go to the store, you are restricted to your Technician privileges.

What if you are the guest and have a higher-class license than the host? A guest operator hosted by a lower-class licensee can use their higher-class privileges as the control operator whether the host is present or not. In this case, there are special identification rules described below.

Regardless of license class, though, both the guest operator and station owner are responsible for proper operation of the station. The control operator is responsible for proper operation. The station owner is responsible for limiting access to the station to only responsible licensees that will follow the FCC rules.

UNLICENSED OPERATORS

Here's a situation that you may already have experienced. Have you visited a club or personal station and made a contact or two? Perhaps you took part in Field Day or a radio contest with licensed hams. Maybe you just took the microphone to say hello to a friend over the air. If so, you have acted as an unlicensed operator under the supervision of a control operator.

There's nothing at all wrong with letting an unlicensed operator use your station—as long as a control operator is present when transmissions are made. Unlicensed operators may not act as control operators, including family members of licensed hams. This is sometimes difficult for family to understand—even though they can probably turn on the rig and use the microphone, they are not allowed to do so because they are not licensed. No license equals no control operator equals no unsupervised transmissions.

It's not OK for a control operator to let family members operate while they're elsewhere on the property, doing a chore or grabbing a snack. The control operator must be present and insuring that all FCC rules are met. One good thing, unlicensed operators can use the equipment to receive at any time!

Before you go on, try questions T1C01; T2C01, 03-06, 12; T2D01-03. If you have difficulty, come back to this section.

6.2 Identification

In most contacts the other party can't see you—they have no other way of identifying your signals besides your call sign. It's also important that other stations be able to determine who is transmitting. Your call sign is your identity on the air and "Identify" or "Identification" means to send or speak your call sign over the air.

Identification rules apply whether you're operating at home, on the road, or floating in a hot-air balloon. Proper identification is important not only so the FCC knows who's transmitting, but for any other station that wants to contact you or to know where your signal is coming from. This is why identification is one of the few areas in which the FCC tells you how to operate!

NORMAL IDENTIFICATION

The first rule of identification is that unidentified transmissions are not allowed, no matter how brief. Unidentified means that no call sign was associated with a transmission. If you need to make a short transmission to test an antenna or radio, just speaking your call sign will suffice. For example, to see if you are in reach of a distant repeater, don't just key the microphone and listen for the repeater's ID or squelch tail—that's called *kerchunking* for the sound that everyone monitoring will have to endure. Just state your call sign once as you transmit. No problem!

The identification rules are simple—give your call sign at least once every ten minutes during a contact and when the contact is finished. What? Not at the beginning of the contact? You generally need to give your call to establish contact, so that's a moot point. What if the contact is too short for the ten minute rule? Just give your call sign as you end the contact.

You don't have to give the other station's call sign, either. The purpose of identification is to identify the source of *your* signal, not those of other stations. Giving the other station's call sign is for convenience and is considered good practice. Here's a tip, it's not necessary to say, "For ID," since whenever you give your call sign, you ID!

Your call can be given in Morse code, by voice, in an image, or as part of a digital transmission. Video and digital call signs will need to be transmitted in a standard protocol or format so that anyone can receive it. If you are communicating in a language other than English, you are required to identify in English. The FCC recommends the use of phonetics (see Chapter 4) when you identify by voice—that avoids confusing letters that sound alike. Morse code, by the way, can be used for identification no matter in what mode the contact is being conducted!

Tactical calls

Tactical call signs (or tactical IDs) are used to help identify where a station is and what it is doing. Examples of tactical calls include "Waypoint 5", "First Aid Station", "Hollywood and Vine", and "Fire Watch on Coldwater Ridge." Tactical calls can be used at any time, but are usually used in emergency and public service operation when providing communications.

Tactical calls don't replace regular call signs and regular identification rules apply—every 10 minutes and at the end of contact. Tactical calls allow consistent identification that streamlines communication based on function. It would be really confusing if everyone had to remember which individual call sign was performing which function. It's common, for example, that a station is manned by different operators at different times. When a new operator takes over, he or she simply gives their FCC-assigned call sign along with a tactical call, such as "This is W1AW at Race Headquarters." They use "Race Headquarters" as the tactical call from then on, giving the regular call sign (W1AW) once every ten minutes.

At this station, Rebecca Day, KS4RX (left) acts as control operator for Amanda, KG4NBF. Rebecca is responsible for making sure all FCC rules are followed.

GUEST OPERATORS

When you are visiting another station, unless your host "lends" you the station, you must identify using the call sign of the host. It doesn't matter who has the highest-class license. If the guest operator has a higher class license than the host, the guest identifies with the call sign of the host followed by their own call sign.

For example, if KD7FYX (a General-class licensee) uses the station of W7FIR (a Technician-class licensee) on the HF bands, he must sign "W7FIR/KD7FYX". On a VHF band, signing "W7FIR" would be sufficient since both licensees have the same privileges on VHF.

This is a little easier to understand if you think about it from the standpoint of a listening station. If they heard W7FIR on 20-meter phone and looked up the call in the FCC data base, they would think W7FIR was using a frequency not allowed for Technician-class licensees. Adding KD7FYX makes it clear whose privileges are being used. Needless to say, identifying is more efficient if W7FIR lends the station to KD7FYX.

MISCELLANEOUS IDENTIFICATION RULES

There are two exceptions to the identification rules and that is for remote control signals and signals retransmitted through space stations. If you are controlling a model aircraft, for example, you don't have to send your call sign. Remote control signals are weak and don't travel long distances, so a call sign is not of much use. You do, however, have to attach your call sign to the transmitter.

Space stations, such as amateur satellites and stations on the International Space Station (ISS) or space shuttle, have special rules for identification. Space stations do not have to identify themselves although the station on the ISS uses the call signs NA1SS and RUØSS. Many amateur satellites retransmit signals from the ground using *transponders* (systems that retransmit an entire range of frequencies on a different band) and so do not add their own call sign to the output.

Test transmissions

Identification rules apply to test transmissions, as well. The call sign must be given once every ten minutes and at the end of transmissions. Test transmissions must be kept brief to avoid causing interference or to keep from occupying an otherwise useful frequency. The usual method of identifying during a test transmission is to say "W1AW testing" or send "W1AW VVV", where "V" is usually used as a test signal on Morse code.

Automatic Identification

Stations under automatic control (see Chapter 6.5) must also identify themselves. This is part of the requirement that automatically-controlled stations be controlled by procedures or devices that make sure FCC rules are followed. Repeaters are the most common type of automatic station. Repeaters identify themselves by transmitting the

Licensee or Trustee?

(From the *ARRL FCC Rule Book*) There seems to be a lot of confusion as to whether the person whose name is on the license for a repeater or a remote base is the "licensee" or the "trustee" of the station. The operator could be either one! It depends upon the type of license the station is operating under.

If the repeater or remote base is operating under the auspices, and using the call sign of an individual amateur's personal station license, then the operator is the "licensee" of the station, not the "trustee."

If it is operating under the auspices of an FCC-issued club station license, and using the FCC-issued club call sign, then the person whose name appears on the license is the "trustee," not the "licensee."

K6LL is an active contester. He knows and practices all the FCC rules that govern Amateur Radio, including, the need to identify your station. The purpose of identification is to identify the source of *your* signal, not those of other stations. Giving the other station's call sign is for convenience and is considered good practice.

station call sign by voice, by Morse code (20 WPM or slower), or as an image in a standard video signal format.

Special Event Stations

If an amateur club with an FCC license arranges to use a special event call sign, such as the popular 1-by-1 calls (see Chapter 5), both the regular call and the special event call must be given on the air. The usual call of the amateur or club requesting the special event call sign must be given once per hour. This allows listeners to determine the identity of the station, since special event calls are short-lived and sometimes reassigned.

Before you go on, try questions T2B01-07, 10; T2A05; T3A05-07. If you have difficulty, return to this section.

Harmful interference is not necessarily intentional; it may simply be due to an overloaded receiver! Modern receivers are tremendously sensitive, but it's asking a lot of them to run at full sensitivity while rejecting strong signals nearby. Gadgetry such as noise blankers and preamplifiers can create problems where there are none. Harmful interference can often be mitigated with good receiver operating technique

The biggest problem of all is the Noise Blanker. Most noise blankers operate by sensing short, sharp noise pulses. They look at an entire band, not just what is coming through the narrow IF filters. A strong nearby signal can confuse a noise blanker to the point of nearly shutting down a receiver or causing what sounds like severe splatter over many kHz. Unless you have really strong local line or ignition noise, turn your noise blanker OFF. If the band is full of strong signals, noise blankers are useless or worse.

The RF Attenuator can be your biggest friend when dealing with strong nearby signals. It's surprisingly easy for a strong signal to drive a receiver into overload. This creates spurious receiver responses, known as "crud," up and down the band. Switching in the attenuator cures a surprising number of ailments when your receiver is no longer being overloaded. Remember that the goal is to maximize signal-to-noise ratio for understandability. Try out your attenuator and you may be surprised at how much it cleans up a band!

The RF Gain control can make your receiver very sensitive but susceptible to overloading. Experiment with reducing RF Gain to see if it doesn't improve your receiver's performance on a busy band. Even during casual operating, backing off the RF Gain can dramatically reduce background noise.

Does your receiver have Passband Tuning, IF Shift, Variable Bandwidth or similar controls? All those new Digital Signal Processing features you paid for can also clean up noise and attenuate low-frequency or high-frequency audio. Find the receiver's manual and learn what these controls do. By effectively using the capabilities of a modern receiver, you will surely find that the band is quieter and nearby signals less disruptive. In fact, you will find yourself making better use of your receiver's controls every day!

6.3 Interference

Interference is caused by "noise" and by "signals." Noise interference is caused by natural sources, such as thunderstorms (atmospheric static is referred to as *QRN*), or by signals unintentionally radiated by appliances, industrial equipment, and computing equipment. The type of interference discussed in this part of the book is caused by signals from other transmitters.

Interference from nearby signals, or *QRM*, is part of the price of frequency flexibility. If hams operated on assigned and evenly-spaced channels, there would be much less interference. The channels would also be frequently overloaded! Interference is not necessarily illegal, just inconvenient. Hams have learned various ways of dealing with QRM starting with the following:
- Common sense and courtesy help avoid many problems
- Be sure to equip your radio with good filters to reject interference
- Remember that no one owns a frequency
- Be aware of other activities, such as special events, DXpeditions, and contests

Most interference is manageable!

HARMFUL INTERFERENCE

If a transmission disturbs other communications, that's considered harmful interference. Every ham should make sure to both transmit and receive in a way that minimizes the possibility of causing harmful interference. Reports of interference such as transmit-

Interference, otherwise known as *QRM*, causes severe distortion to this image sent by Amateur Radio "slow scan" TV. While harmful interference can be vexing, accidental interference is common, as is the case here.

ting off-frequency or generating spurious signals (splatter and buckshot) should be checked out. When testing or tuning a transmitter, use a dummy load and always keep your test transmissions short. Use only the power necessary to make the contact.

While harmful interference can be vexing, accidental interference is common. For example, propagation on a band can change due to ionospheric or atmospheric conditions. A signal that wasn't there a few minutes ago might suddenly become strong enough to disrupt your contact. Changing an antenna direction can allow a previously rejected signal to appear or transmit a signal towards other stations. Sometimes, a station will start listening during a pause in activity and start transmitting thinking the frequency is available. These things happen and one can't demand a perfectly clear frequency.

What should you do if harmful interference occurs to your contact? Assuming the interfering station isn't intentionally trying to cause interference, can you change frequency a little bit or change antenna direction? Be flexible—it's one of ham radio's greatest strengths! What should you do when you cause harmful interference? If it's your fault, apologize, identify, and take the necessary steps to reduce interference—change frequency, reduce power, or move your antenna.

WILLFUL INTERFERENCE

Intentionally creating harmful interference is willful interference and is never allowed. The interference doesn't have to be aimed at one specific contact or group. Any time communications are deliberately disrupted, that's willful interference. For example, intentionally transmitting spurious signals by overmodulating is willful interference. Luckily, willful interference is pretty rare on the ham bands since most people have the good sense and maturity to not do it.

Before you go on, try questions T1A10; T3D01, 04, 05, 08. If you have difficulty, return to this section.

Table 6-1
Third-party Agreements
The United States has third-party agreements with the following nations...

V2	Antigua/Barbuda	6Y	Jamaica
LU	Argentina	JY	Jordan
VK	Australia	EL	Liberia
V3	Belize	V7	Marshall Islands
CP	Bolivia	XE	Mexico
T9	Bosnia-Herzegovina	V6	Micronesia, Federated States of
PY	Brazil	YN	Nicaragua
VE	Canada	HP	Panama
CE	Chile	ZP	Paraguay
HK	Colombia	OA	Peru
D6	Comoros (Federal Islamic Republic of)	DU	Philippines
TI	Costa Rica	VR6	Pitcairn Island
CO	Cuba	V4	St. Christopher/Nevis
HI	Dominican Republic	J6	St. Lucia
J7	Dominica	J8	St. Vincent and the Grenadines
HC	Ecuador	9L	Sierra Leone
YS	El Salvador	ZS	South Africa
C5	Gambia, The	3DA	Swaziland
9G	Ghana	9Y	Trinidad/Tobago
J3	Grenada	TA	Turkey
TG	Guatemala	GB	United Kingdom
8R	Guyana	CX	Uruguay
HH	Haiti	YV	Venezuela
HR	Honduras	4U1ITU	ITU-Geneva
4X	Israel	4U1VIC	VIC-Vienna

Third party communication in action! "Miss Mississippi" Jalin Wood, who is not a ham, sends CW with some help from K3QC.

6.4 Third-party Communications

Ham radio is frequently used to send messages, written or not, on behalf of unlicensed persons or organizations. One of the oldest activities in ham radio is the sending of messages, relaying them from station to station until delivered by a ham near the addressee. This is *third-party communication*. Because third-party communications bypass the normal telephone and postal systems, many foreign governments have an interest in limiting it. Looking at third-party communications from the ham radio side, the FCC does not want the Amateur service to become a non-commercial messaging system. So, we have some rules specifically governing third-party communications.

DEFINITIONS & RULES

Let's start by defining the important aspects of third-party communications:

● The entity on whose behalf the message is sent is the "third party" and the control operators that make the radio contact are the first and second parties. A third party can also be the recipient of a message generated by a ham.

- A licensed amateur capable of being a control operator at either station is not considered a third party.
- The third party need not be present in either station. A message can be taken to a ham station or a ham can transmit speech from a third-party's telephone call over the ham radio—this is called a *phone patch*.
- The communications transmitted on behalf of the third party need not be written. Spoken words, data, or images can all be third-party communications.
- The third party may participate in transmitting or receiving the message at either station. An unlicensed person in your station sends third-party communications when they speak into the microphone, send Morse code, or type on a keyboard.
- An organization, such as a church or school, can also be third parties.

Simplifying the definition, any time that you send or receive information via ham radio on behalf of any unlicensed person or an organization, even if the person is right there in the station with you—that's third-party communications.

The FCC recognizes that third-party communications is a vital part of ham radio and its mission, specifically to train operators and to provide an effective emergency communications resource. Handling messages, or *traffic*, phone patches, and live conversations are all part of both normal and emergency communications. As a result, third-party communications may be exchanged between any amateur stations operating under FCC rules with the constraint that the communications must be noncommercial and of a personal, nature.

When signals cross borders, the rules change. International third-party communications are restricted to those countries that specifically allow third-party communications with US hams. **Table 6-1** shows what countries have *third-party agreements* with the United States. If the other country isn't on this list, third-party communications to or from that country is not permitted.

This is all much clearer if illustrated by some examples:

- A message from one ham to another ham is not third-party communications, whether directly transmitted or relayed by other stations.
- Letting an unlicensed neighbor make a contact under your supervision is third-party communications, even if the contact is short and for demonstration or training purposes.
- If you contact a DX station who asks you to pass a message to his family in your state, doing so would be third-party communications. Check to be sure the DX station's country has a third-party agreement with the US before accepting the message.
- Making a contact to allow a visiting student to talk to his family in South America is third-party communications even if both the student and the family are present at the stations involved. Be sure there is a third-party agreement in place.

Before you go on, try question T2D05. If you have difficulty, come back to this section.

6.5 Remote & Automatic operation

Many stations, such as repeaters and beacons, operate without a human control operator present to perform control functions. It is also becoming common to operate a station via a link over the Internet or phone lines. These two types of operation are specially defined in the rules, but the bottom line remains the same—the station must be operated in compliance with FCC rules, no matter where the control point is located.

DEFINITIONS

Local control - a control operator is physically present at the control point. This is the situation for nearly all amateur stations, including mobile operation. Any station can be locally controlled.

Remote operation - the control point is located away from the transmitter, but a control operator is present at the control point. The control point and the transmitter are connected by some kind of control link; by Internet, phone line, or radio. The numerous stations that operate using an Internet link are good examples of remotely operated stations. Any station can be remotely controlled, as well.

Automatic operation - the station operates completely under the control of devices and procedures that insure compliance with FCC rules. Repeaters, beacons, and space stations can be automatically controlled.

RESPONSIBILITIES

No matter what type of control is asserted—local, remote, or automatic— the station must operate in compliance with FCC rules at all times. No excuses! Automatic stations might not have a control operator controlling the station at all times, but a control operator must be responsible for the station's operation.

Repeater owners must install the necessary equipment and procedures for automatic control that ensure the repeater operates in compliance with FCC rules. If automatic control results in rule violations, The FCC can require a repeater to be placed on remote control (meaning that a control operator must be present when the repeater is operating). Repeater users are responsible for proper operation via the repeater, however.

Because digital protocols are designed to operate automatically, there are special rules for automatic control when using them. Stations using a data mode (including RTTY) may operate under automatic control in certain segments of the HF bands and above 50 MHz as listed in rule [97.221] (b). Data stations are the only type of automatically-controlled station allowed to forward third-party communications. (Note that it is OK to pass third-party messages using a repeater.)

> *Before you go on, try questions T2C05, 07-11. If you have difficulty, return to this section.*

6.6 Prohibited Transmissions

Not many types of transmissions are specifically prohibited because amateurs are given wide latitude to communicate within the technical and procedural rules. While other services are very tightly regulated, hams are encouraged to experiment and be flexible. There are limits—here are four types of communications prohibited for hopefully obvious reasons:

- Unidentified transmissions - any transmission without an identifying call sign during the required time period.
- False or Deceptive Signals - transmissions intended to deceive the listener, such as using someone else's call sign.
- False distress or emergency signals - because of the legal requirement to respond, these are taken very seriously by the FCC and other authorities.
- Obscene or Indecent Speech - avoid controversial topics and expletives. There is a more complete discussion in the *ARRL's FCC Rule Book*.

Generally speaking, regular communications that could reasonably be performed through some other radio service are also prohibited. For example, regularly using ham radio to direct boat traffic on a lake should be done on marine VHF channels. Communications in return for some kind of compensation is also prohibited as we discuss below. None of these prohibitions are unreasonable and help keep Amateur Radio a useful and rewarding activity free of commercial intrusion.

BUSINESS AND EMPLOYER COMMUNICATIONS

No transmissions related to conducting your business or employer's activities are permitted. This is, after all, *Amateur* Radio, and there are plenty of radio services available for commercial activities. However, one's own personal activities don't count as "business" communications. For example, it's perfectly OK for you to use ham radio to talk to your spouse about doing some shopping or to confer about what to pick up at the store. You can even order things over the air, as long as you don't do it regularly and as part of your normal income-making activities. Here are some examples of acceptable and non-acceptable activities:

OK
- Using a repeater's autopatch to make or change a doctor's appointment
- Advertising a radio on a swap-and-shop net
- Describing your business as part of a casual conversation

NOT OK
- Using a repeater's autopatch to call a business client and change an appointment
- Selling household or sporting goods on a swap-and-shop net
- Regularly selling radio equipment at a profit over the air
- Advertising your professional services over the air

One exception to the profession or business prohibition is that teachers may use ham radio as part of their classroom instruction. In that case, they can be a control operator of a ham station, but it must be incidental to their job and can't be the majority of their duties.

Another broad prohibition is being paid for operating an amateur station. Your employer can set up an emergency amateur station, and even pay you to build it, but you can't be paid for time you spend operating it. This is also true for employees of public safety and medical organizations. While it's a great idea to have emergency stations in hospitals and fire stations, hospital staff and firefighters can't use the station during their paid time at work.

An exception to the no-paid-operators rule is for club stations. They are allowed to hire an operator to send bulletins and information of general interest to hams. For example, it's OK for W1AW (the ARRL club station) to employ operators to send out bulletins, but they can't engage in ragchew contacts afterwards while on the ARRL clock. The operators can be conpensated (paid) only if station transmits 40 hours of bulletins and/or code practice per week.

ENCRYPTED TRANSMISSIONS

Amateur Radio is a public form of communication. That is part of the trade-off for being licensed to use all of our operating privileges. As a consequence, it is not OK to transmit secret codes or to obscure the content of the transmission if the intent is to prevent others from receiving the information.

Translating information into data for transmission is called *encoding*. Recovering the encoded information is called *decoding*. Most forms of encoding are OK because they are done according to a published digital protocol. Any ham can look up the protocol and develop the appropriate capabilities to receive and decode data sent with that protocol.

Encoding designed to prevent the information from being decoded is called *encryption*. Recovering the encrypted information is called *decryption*. Amateurs may not use encryption techniques except for radio control and control transmissions to space stations where interception or unauthorized transmissions could have serious consequences.

The difference between encoding and encryption are sometimes not always clear-cut. If a transmission is encoded according to an obscure and little-used protocol, for most hams it might as well be encrypted. But as long as the protocol is published and available to the public, that transmission is acceptable. The general rule to remember is that no ham should be prevented from receiving the communications of another ham by withholding the necessary information.

BROADCASTING

Non-hams often refer to ham transmissions as "broadcasting" but that is inaccurate. Broadcasting consists of *one-way transmissions* intended for reception by the general public. Hams are not permitted to make this type of transmission. The prohibition on broadcasting includes repeating and relaying transmissions from other communications services. Hams are also specifically prohibited from assisting and participating in news gathering by broadcasting organizations. Code practice and bulletins for amateurs are OK, but news bulletins are not.

The prohibition against transmission of music (and other entertainment-type material in video and image transmissions) extends to incidental retransmission of music from a nearby radio. This means that you should turn down the car radio or CD player when you're using the ham radio! Music can only be rebroadcast as part of an authorized rebroadcast of space station or space shuttle transmissions—a rather unusual circumstance.

SPECIAL CIRCUMSTANCES

Ham communications must be intended for reception by hams. This leads to some exceptions from the normal broadcasting rules—hams may retransmit weather and propagation information from government stations, but not on a regular basis. As always, in an emergency, hams are authorized to use whatever means necessary to prevent loss of life or property. When the emergency passes, however, regular rules return.

Hams like to operate from unusual places and while in motion, so what about operating from a plane or boat? These circumstances are covered in rule 97.11. You may operate, but only with the approval of the captain. (All commercial passenger flights have strict bans on transmitting from inside the aircraft while in flight.) In the case where you do get permission, such as on a private plane, you have to use your own radio equipment and can't use any of the aircraft's or boat's radio equipment. Amateur transmissions may not interfere with any of the other radio systems on-board, including aircraft used in the boat's operation, such as helicopters, and navigation systems.

In general, hams can't communicate with non-amateur services, but the FCC may allow hams to talk to non-ham services at certain times or during a declared communications emergency. RACES operators may communicate with government stations during emergencies, too. And for fun, once a year, the FCC permits ham-to-military communication on Armed Forces Day during May—see *QST* magazine or the ARRL's Special Events Web calendar at **www.arrl.org/contests/spev.html**.

Before you go on, try questions T2A01-4, 6-11; T1C03, 11; T2D04, 07, 11. If you have difficulty, return to this section.

Chapter 7

Electrical and RF Safety

In this section, you'll learn about:

- Electrical safety
- Reducing electrical hazards
- Working safely with electricity
- Grounding for safety
- Grounding at RF
- Lightning protection
- RF exposure rules
- Evaluating exposure from your station
- Reducing exposure to RF
- Automotive station safety
- Safely installing antennas
- Working around towers

There is nothing particularly risky about working with electricity, even though you can't see it. Compared to many activities, radio is one of the safest hobbies for people of all ages. Most hams go through an entire lifetime of ham radio without having a serious safety incident. This is at least partially due to educating themselves about safety and then following simple rules.

Safety is just as important for radio as it is for house wiring or working on an engine or using power tools. The keys to safety are understanding the potential hazards, mitigating them, and being able to respond to an injury in the unlikely event that one occurs. By being informed and prepared, your exposure to electrical hazards is greatly reduced.

Depending on your background, some of this material will be a review—nothing wrong with learning safety twice! Other topics specifically about radio will likely be new. In either case, safety is important enough that you'll encounter several questions about it on your exam. We'll begin with basic electrical safety information before moving on to radio. Safety coverage concludes by reviewing some of the mechanical aspects of your radio activities.

Remember, this information is not here to frighten you. Radio and electricity are not intrinsically unsafe. You are probably being educated about radio as an unfamiliar technology. Wouldn't it be a good idea to learn the appropriate safety techniques at the same time? Of course it would!

7.1 Electrical safety

Working safely with electricity mostly means avoiding contact with it! Most modern ham radio equipment is solid-state but the ac line voltage that powers most equipment is dangerous. You may also encounter vacuum tubes and the high voltages they use in power amplifiers. Treat electricity with respect.

ELECTRICAL INJURIES

Electrical hazards can result in two types of injury; shocks and burns. Whenever electricity can flow through any part of your body, both can occur. Shocks and burns can be caused by ac or dc current and result from current flow through the body.

Depending on the voltages present, shocks and burns can range from insignificant to deadly. Voltage is what causes the current to flow, but doesn't shock all by itself. Just like in a regular resistor, as the voltage applied across your body varies, so does current. While parts of the body like hair and fingernails are not good conductors, the interior of the body

Table 7-1

Effects of Electric Current Through the Body of an Average Person

Current (1 Second Contact)	Effect
1 mA	Just Perceptible.
5 mA	Maximum harmless current.
10 - 20 mA	Lower limit for sustained muscular contractions.
30 - 50 mA	Pain
50 mA	Pain, possible fainting. "Can't let go" current.
100 - 300 mA	Normal heart rhythm disrupted. Electrocution if sustained current.
6 A	Sustained heart contractions. Burns if current density is high.

conducts electricity quite well, being mostly salty water.

Electrical currents of more than a few mA can cause involuntary muscle spasms, which leads to the jerking and jumping image imagined on TV and in the movies. No joking matter, muscle spasms can cause falls and sudden large movements. The sudden pulling back of an outstretched hand or finger that comes in contact with an energized conductor is a result of arm muscles contracting. **Table 7-1** lists some of the effects of current in the body.

While any shock can be painful, the most dangerous currents are those that travel through the heart, such as from arm-to-arm or arm-to-foot. Electrical currents of 100 mA or more can disrupt normal heart rhythm. Depending on the resistance of the path taken by the current, voltages as low as 30 volts can cause enough current flow to be dangerous.

Burns caused by dc current or low-frequency ac current are a result of resistance to current in the skin, either through it to the body's interior or along it from point to point. The current causes heat and that's what results in the burn. At higher frequencies, an *RF burn* can result from contact (or near contact) with a "hot spot"—a location where high RF voltage is present on the outside of a connector, cable, or equipment enclosure. RF burns are caused by the RF voltage causing shallow currents in the skin at the point of contact. While they can be painful, RF burns generally don't do much damage. Larger-scale burns from dc and low-frequency ac are generally deeper and more damaging.

Mitigating Electrical Hazards

Shocks and burns are completely preventable if there is no way for you to come in contact with an energized conductor—simply prevent current flow through the body! Start by never working on "live" equipment unless it is necessary for troubleshooting or testing. Remove, insulate, or otherwise secure loose wires and cables before testing or repairing equipment. Never assume equipment is off or de-energized before beginning your work. Check with a meter or tester first.

If you do need to work on equipment with the power on—sometimes there's no way around it—follow these simple safety steps:

- Remove unnecessary jewelry from your hands because metal is a great conductor. Rings can also absorb RF energy and get quite hot in a strong RF field, such as inside an amplifier, filter, or antenna tuner.
- Keep one hand in your pocket while probing or testing energized equipment and wear insulating shoes. This gives current nowhere to flow in or along your body.
- It's also easy to fall into bad habits after working with low-voltage or battery-powered equipment that poses few hazards. Be extra careful when changing to work around higher-voltages.
- Never bypass a safety interlock during testing unless specifically instructed to do so. Safety interlocks remove power when access panels, covers, or doors are opened to hazardous areas in equipment. They are used to prevent unintentionally opening a cabinet or enclosure where dangerous voltages or intense RF may be present.
- Capacitors can store charge after a charging circuit is turned off, presenting a

hazardous voltage for a long time. This includes small-value capacitors charged to high-voltage! Make sure all high-voltage capacitors are discharged by testing them with a meter or use a *grounding stick* to shunt their charge to ground.

- Storage batteries release a lot of energy if shorted. Keep metal objects such as tools and sheet metal clear of battery terminals and avoid working on equipment with the battery connected.
- Avoid working alone around energized equipment.
- Remember that electricity can move a *lot* faster than you! Even your quickest touch is plenty long enough for electricity to flow.

The *ARRL Handbook* has an excellent discussion of workbench and radio shack safety. You can also follow the links on the *Ham Radio License Manual* Web page and the ARRL's Technical Information Service Web page (see **www.arrl.org/hrlm**) to more articles on all kinds of radio and electrical safety, including first aid for electrical injuries.

Response to Electrical Injury

If you were shocked in your radio shack, would others be able to help? The first step in any first aid response to electrical injury is to remove power. Install a clearly-labeled master ON/OFF switch for ac power to your station and workbench. It should be located away from the electrically-powered equipment where you are likely to be. Show your family and friends how to turn off power at the master switch and at your home's circuit-breaker box.

It's also a good idea for all sorts of reasons for you and your family to get CPR training and to learn how to administer first aid for electrical injuries. To learn more about responding to electrical injuries browse the Healthopedia's entries on "electrical injury" at **www.healthopedia.com/electrical-injury/treatment.html**.

WIRING AND SAFETY GROUNDING

A large part of electrical safety is to not create hazards in the first place! This is why the National Electrical Code and your local building codes were created—to prevent common electrical hazards. A home wired "to code" has properly sized outlets and wiring and a *safety ground* to help prevent shocks. The safety ground is a connection to a ground rod at your home near the main electrical service entry. It provides a path to ground for current in case of an accidental short-circuit between either the hot or neutral wires and an appliance's metal enclosure or *chassis*.

Most ham stations don't require new wiring and can operate with complete safety when powered from your home's ac wiring. That is, as long as you follow simple guidelines:

- Use three-wire power cords that connect the chassis of your equipment to the ac safety ground.
- Use Ground Fault Interrupter (GFI) circuit breakers or circuit breaker outlets.
- Verify ac wiring is done properly by using an ac circuit tester.
- Never replace a fuse or circuit breaker with one of a larger size.
- Don't overload single outlets.

If you do decide to run new wiring for your station as it grows, either have a licensed electrician do the wiring or inspect it. Be sure to follow the hot-black (occasionally red)/neutral-white/ground-green or bare wiring standard shown in **Figure 7-1**. Use cable and wire sufficiently rated for the expected current load as

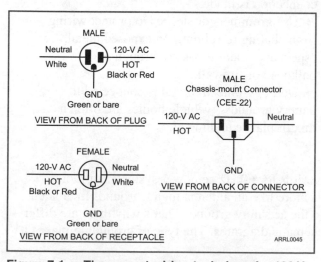

Figure 7-1 — The correct wiring technique for 120 V ac power cords and receptacles. The white wire is neutral and the green wire is the safety ground. The hot wire can be either black or red. These receptacles are shown from the back, or wiring side.

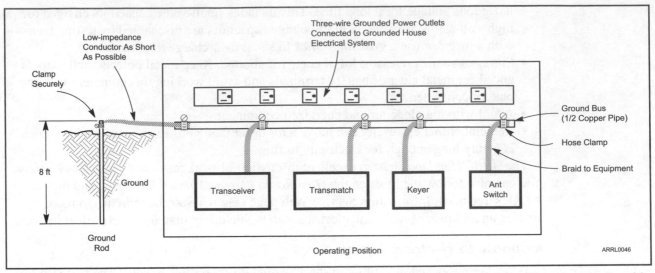

Figure 7-2—An effective station RF ground consists of keeping all equipment bonded together electrically with a short, low-impedance connections using a single common bus, such as copper pipe, flashing, or even coaxial cable braid. The bus is then connected to a nearby ground rod with braid or strap.

Table 7-2

Current-Carrying Capability of Some Common Wire Sizes

Copper Wire Size (AWG)	Allowable Ampacity (A)	Max Fuse or Circuit Breaker (A)
6	55	50
8	40	40
10	30	30
12	25 (20)[1]	20
14	20 (15)[1]	15

[1]The National Electrical Code limits the fuse or circuit breaker size (and as such, the maximum allowable circuit load current) to 15 A for #14 AWG copper wire and to 20 A for #12 AWG copper wire conductors.

shown in **Table 7-2**. Use the proper size fuses and circuit breakers.

Mobile installations have special requirements for safe wiring. Remember that a vehicle battery stores a lot of energy! Accidental short circuits can not only damage your radio equipment, they can start fires and do a lot of expensive damage to your car. General guidelines include:

- Fuse both the positive and negative leads of your radio.
- Connect the radio's negative lead to the vehicle body where the battery ground connection is made.
- Use grommets or sleeves to protect wiring from chafing or rubbing on exposed metal, especially where it passes through a bulkhead or firewall.
- Don't assume all metal is connected to battery ground— vehicle bodies are often a mix of plastic and metal.

RF GROUNDING

A home's safety ground is adequate to control shock hazards for 60 Hz ac power systems. At radio frequencies, though, the safety ground wiring usually acts more like an antenna than a ground! In amateur stations, it's necessary to provide a separate *RF ground* to the ac safety ground. That's why there are different symbols for ground as you found in the section on schematic diagrams. The type of grounding makes a difference in radio systems.

RF grounding prevents problems associated with RF currents in the station. Using a common ground keeps all of the radio equipment at the *same* RF voltage—even if it's not exactly at ground potential. RF current in your station can cause audio distortion, erratic operation of computer equipment, and occasionally RF "burns" as noted earlier. This is because the feed line, equipment enclosures, and the connections be-

To provide both an additional power wiring ground and an RF ground, many hams use a ground rod outside of their house near the shack. A ground rod should be at least 8-feet long—it may take more than one shorter rod if you are unable to install a full-length ground rod. The ground rod should be either solid copper (heavy-wall pipe is fine) or copper-clad steel (easier to drive in). Beware of thinly-plated ground rods because their copper will quickly erode, leaving a poor ground of rusty steel.

To act as an RF ground on the HF bands, it should be as close as possible to your equipment. Lengths of 8 to 10 feet or more start to act like antennas on the higher HF bands, such as 12 and 10 meters. At VHF and UHF, the purpose of the ground rod is primarily a safety ground.

From the ground rod, run a very heavy solid wire (8 gauge or larger) or a strip of copper flashing to the ground bus near your equipment. Use a standard ground rod clamp to attach the ground connection and be sure to use anti-corrosion compound. Solid wire is best because it presents the least impedance to RF current. Braid from coaxial cable such as RG-8 or RG-213 is also acceptable.

If you live in an apartment or have a shack above the ground floor, an external ground rod may not be practical. The length of connection from the equipment would probably be too long to act as an RF ground on any band. The same holds true for cold-water pipes that travel for long distances inside the building. In such cases, it is best to just use a ground bus near your equipment to keep it all at the same RF voltage and prevent RF current from flowing between individual pieces of equipment. Add a connection to the ac safety ground for electrical safety.

Basement or ground floor stations may be able to use a cold water pipe for a ground, since they travel underground for many feet, connecting to the water main. (Hot water pipes have to go to a water heater and so do not generally make good RF grounds.) Before you decide to use a pipe, make sure it takes a short, direct route to exit the house. It doesn't do much good if the pipe runs all the way across the house before entering the ground! Also, make sure the pipe is metallic—plastic pipe is very common and is worthless as an RF ground.

tween them are acting as antennas for your transmitted signal. The current flowing in sensitive audio cables or data cables can interfere with their normal function, just as your strong transmitter signal might be picked up and detected by a neighbor's telephone or audio system. If all of the equipment surfaces are kept close to the same RF voltage, no RF current will flow between them, eliminating the unwanted effects.

Every station is different and so will have to create a unique RF ground, but **Figure 7-2** shows the general idea.

- Bond all metal equipment enclosures to a common ground *bus*.
- Keep all connections, straps, and wires short to keep ground connections from acting like antennas.
- Connect the ground bus to a ground rod or grounded pipe with a short, wide conductor such as copper flashing or strip.
- In difficult situations, a piece of wide flashing or screen can be placed under the equipment and connected to the ground bus.

The ARRL Technical Information Service (**www.arrl.org/tis/info**) dedicates an entire page to RF grounding with several *QST* magazine articles and Web references. Each station is a little different and you may have to experiment to get the results you expect. Low power VHF/UHF stations usually have few RF grounding problems.

LIGHTNING

Even though amateur antennas and towers are struck no more frequently than tall trees or other nearby structures, it is wise to take some precautionary steps. This is especially true for stations in areas with frequent severe weather and lightning.

Starting at your antennas, all towers, masts, and antenna mounts should be grounded. This can be done at their base, or in the case of roof mounts, though a heavy wire to a ground rod. Provide a common entry panel where cables enter the house and install lightning arrestors at the entry panel. A ground rod should be nearby, if this is not the location of the station ground rod. *The ARRL Ham Radio License Manual's* Web page (**www.arrl.org/hrlm**) lists links to resources on lightning protection.

When lightning is anticipated, the best protection is to disconnect all cables outside the house and unplug equipment power cords inside the house. This interrupts the lightning's path to get to ground. In fact, it's not a bad idea to disconnect both power and phone cords to household appliances and long network and signal cables to your computing equipment. If you think you will unplug your equipment frequently, it might be a good idea to use or make power strips so that you can unplug many pieces of equipment with a single cable. Don't just turn them off—lightning jumps across switches quite easily—physically unplug the power cable.

Determine whether your renter's or homeowner's insurance will cover you for damage from a lightning strike and whether the presence of antennas modifies that coverage. Your insurance agent will be able to help you determine the exact coverage and whether any special riders or amendments are needed. You may want to investigate the equipment insurance available to ARRL members, as well.

Regardless of how much protection you install on your antenna system, operating during a thunderstorm is a bad idea. Even a nearby strike can create a voltage surge of thousands of volts in a power or phone line causing equipment damage or setting the house on fire!

> *Before you go on, study questions T0A01-03, 06-08, 12, 13. If you have difficulty, return to this section and try again.*

7.2 RF Exposure

In recent years, there has been a lot of discussion about whether there are health and safety hazards from exposure to electromagnetic radiation (EMR). Many studies have been done at both power line frequencies (50 and 60 Hz) and RF (both short-wave and mobile phone frequencies). No link has been established between exposure to low-level EMR and health risks, including those frequencies used by amateurs. As discussed earlier RF radiation is not the same as ionizing radiation from radioactivity because radio frequencies are not nearly high enough to cause an electron to leave the atom (ionize). At these relatively low frequencies, RF energy is *non-ionizing radiation*.

Nevertheless, even in the absence of evidence that RF fields pose a health risk, it is prudent to avoid unnecessary exposure to high levels of RF. The FCC regulations set limits on the *Maximum Permissible Exposure* (MPE) from radio transmitters of any sort. To abide by these rules without requiring expensive testing, hams are expected to evaluate their stations to see if their operation has the potential to exceed MPE levels. The evaluation process is covered later in this section. The ARRL's *RF Exposure and You* contains a detailed treatment of RF exposure rules and safety techniques.

The only demonstrated hazards from exposure to RF energy are *thermal effects* (heating). *Biological* (athermal) effects have not been demonstrated at amateur frequencies and power levels. Measurable heating occurs only for very strong fields or in fields that originate very close to the body. RF safety techniques involve making sure that persons are not exposed to high-strength fields in one of two ways:

• Preventing access to locations where strong fields are present.

• Making sure that strong fields are not created in or directed to areas where people might be present.

RF burns caused by touching or coming close to conducting surfaces with a high RF voltage present are also an effect of heating. While these are sometimes painful, they are rarely hazardous. RF burns can be eliminated by proper grounding techniques.

Heating as a result of exposure to RF fields is caused by the body absorbing RF energy. Absorption occurs because the RF energy causes the molecules to vibrate at the same frequency. The energy of the vibrations is dissipated within the body as heat. The stronger the field, the more the molecules vibrate and the more heating of the body's tissues results. Absorption also varies with frequency because the body's response also changes with frequency. The total amount of heating then depends on both the RF field's intensity and frequency and is called the *specific absorption rate* or SAR.

Power Density

The intensity of an RF field is called *power density*. Power density, which is an amount of energy per unit of area, can be measured in various ways, but the most common is in milliwatts per square centimeter (mW/cm^2). The power density of an RF field is highest near antennas and in the directions where antennas have the most gain. Power density can also be very high inside transmitting equipment.

Increasing transmitter power increases power density everywhere around an antenna to the same degree that transmitter power increased. Increasing distance from an antenna lowers power density in proportion to the square of the distance from the antenna. i.e.—at twice the distance from an antenna power density is lowered to one-quarter of the value. Controlling these two factors, power and distance, forms the basis for amateur RF safety.

EXPOSURE LIMITS

Safe levels of exposure based on demonstrated hazards have been established by the FCC. These are the Maximum Permissible Exposure levels. Because the SAR varies with

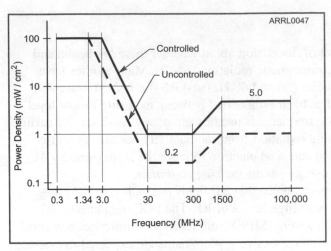

Figure 7-3 — Maximum Permissible Exposure limits vary with frequency because the body responds differently to energy at different frequencies. The controlled and uncontrolled limits refer to the environment in which people are exposed to the RF energy.

Table 7-3
Maximum Permissible Exposure (MPE) Limits

Controlled Exposure (6-Minute Average)

Frequency Range (MHz)	Power Density (mW/cm^2)
0.3-3.0	(100)*
3.0-30	(900/f^2)*
30-300	1.0
300-1500	f/300
1500-100,000	5

Uncontrolled Exposure (30-Minute Average)

Frequency Range (MHz)	Magnetic Field Power Density (mW/cm^2)
0.3-1.34	(100)*
1.34-30	(180/f^2)*
30-300	0.2
300-1500	f/1500
1500-100,000	1.0

f = frequency in MHz
* = Plane-wave equivalent power density

frequency, so does the MPE as shown in **Figure 7-3** and **Table 7-3**. SAR depends on the size of the body or body part affected and is highest where the body and body parts are naturally resonant. The full body is resonant at about 35 MHz if the person is grounded and 70 MHz if they are not grounded. Body parts are resonant at higher frequencies (smaller wavelengths). For example, an adult's head is resonant at the much higher frequency of 400 MHz.

Above and below the ranges of highest absorption, the body responds less and less to the RF energy, just like an antenna responds poorly to signals away from its natural resonant frequency. Frequencies with highest SAR are from 30 to 1500 MHz and these are the regions on the MPE graph where the limits for exposure are the lowest.

Controlled & Uncontrolled Environments

You'll notice on Figure 7-3 that there are two sets of lines; one called *controlled* and the other *uncontrolled*. These refer to the type of environments in which the exposure to RF fields takes place.

• People in controlled environment are aware of their exposure and can take the necessary steps to minimize it.

• People in uncontrolled environments are not aware of their exposure, such as areas open to the general public and your neighbor's property.

The FCC has determined that the higher controlled-environment limits generally apply to amateur operators and members and guests in their immediate households, provided that they are aware of RF fields being used. If this is the case, the controlled-environment limits apply to your home and property—wherever you control physical access

AVERAGING & DUTY CYCLE

Since the effects from RF exposure are related to heating and take place over many seconds, the MPE limits are based on *averages*, not *peak* exposure. This allows exposure to be averaged over fixed time intervals.

• The averaging period is 6 minutes for controlled environments.

• The averaging period is 30 minutes for uncontrolled environments.

The difference in averaging periods reflects the difference in how long people are expected to be

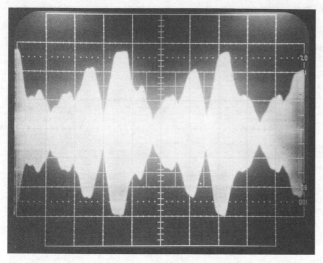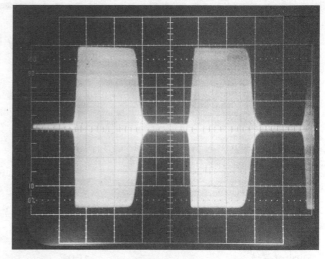

Figure 7-4 — The SSB signal on the left and the Morse signal on the right both have the same peak power, but the average power of the SSB signal is lower.

present and exposed. People are assumed less likely to stay in an uncontrolled environment, receiving continuous exposure, so the averaging period is much longer.

During the averaging period, a transmitter may only be generating RF for a fraction of the time. For most amateur contacts, the transmitter is keyed no more than 50% of the time and usually much less. This pattern lowers the *duty cycle* of the emissions. Duty cycle is the ratio of transmitter on time to the total measurement period and has a maximum of 100%. (*Duty factor* is the same as duty cycle expressed as a fraction, instead of percent, such as 0.25 instead of 25 %.) Because most amateur operation is intermittent, the operating duty cycle of amateur emissions is low, reducing average exposure. For example, during a round-table contact between three stations, each is likely to be talking only one third of the time, for a 33% duty cycle.

Further reducing exposure, some modes have lower average power than others. For example, while sending Morse, the transmitter is off between the dits and dashes. SSB signals only reach peak power for short periods on voice peak and have the lowest duty cycle. FM, however, is a constant power mode and so the signal is continuously at full power when the transmitter is keyed. This is the *emission duty cycle*. For a given PEP, an emission with a lower duty cycle produces less RF exposure as shown in **Figure 7-4**. **Table 7-4** shows the commonly accepted *operating duty cycles* for different modes used by amateurs. PEP multiplied by the mode's operating duty cycle gives the average power for the mode.

To find the average power during the averaging period, use the following method:

Average power = PEP x operating duty cycle x (time transmitting / averaging period)

For example, let's say that your 100-watt transmitter was generating conversational SSB without speech processing. Table 7-4 shows an operating duty cycle of 20% for that mode. During your operating period you transmitted for 1 minute out of every three. Your average power during the evaluation period is:

100 watts x 20% for conversational SSB x (1 min / 1+2 min) = 100 x 0.2 x 0.33 = 6.33 watts

During a 2-meter net as net control, using your 50-watt VHF FM transmitter, you transmit and listen for equal periods. Your average power is:

50 watts x 100% for FM x 50% on/off = 25 watts

Table 7-4

Operating Duty Factor of Modes Commonly Used by Amateurs

Mode	Duty Cycle	Notes
Conversational SSB	20%	1
Conversational SSB	40%	2
SSB AFSK	100%	
SSB SSTV	100%	
Voice AM, 50% modulation	50%	3
Voice AM, 100% modulation	25%	
Voice AM, no modulation	100%	
Voice FM	100%	
Digital FM	100%	
ATV, video portion, image	60%	
ATV, video portion, black screen	80%	
Conversational CW	40%	
Carrier	100%	4

Note 1: Includes voice characteristics and syllabic duty factor. No speech processing.

Note 2: Includes voice characteristics and syllabic duty factor. Heavy speech processor employed.

Note 3: Full-carrier, double-sideband modulation, referenced to PEP. Typical for voice speech. Can range from 25% to 100%, depending on modulation.

Note 4: A full carrier is commonly used for tune-up purposes.

Effect of antenna gain

There is one additional effect that has to be taken into account. As you've learned, beam antennas focus radiated power towards one direction, creating gain. Gain has the effect of increasing your average power in the preferred direction. (It also decreases your average power in other directions.) This means that there are four factors that affect RF exposure; transmitter power and frequency, distance to the antenna, and the radiation pattern of the antenna.

If you use an antenna with gain, you will need to include the effect of gain in your exposure evaluations. For example, if your antenna has 6 dBi of gain, corresponding to a four-fold increase in power radiated in the preferred direction, you would multiply your average power by four.

EVALUATING EXPOSURE

According to FCC rules, all fixed stations must perform an exposure evaluation. (Mobile and handheld transceivers are exempt.) There are three ways of making this evaluation. You could obtain RF power density instrumentation and actually measure the power density of your transmissions. It is also acceptable to make computer models of your station and use those results. Both of these methods are rarely used due to the expense or effort required. By far the most common evaluation used the techniques outlined in the FCC's OET Bulletin 65 (OET stands for *Office of Engineering Technology*). This method uses tables and simple formulas to evaluate whether your station has the potential of causing an exposure hazard. Once you've done an evaluation, you don't

Table 7-5
Power Thresholds for RF Exposure Evaluation

Band	Power (W)
160 meters	500
80	500
40	500
30	425
20	225
17	125
15	100
12	75
10	50
6	50
2	50
1.25	50
70 cm	70
33	150
23	200
13	250
SHF (all bands)	250
EHF (all bands)	250

need to re-evaluate unless you make a significant change in your station, such as increasing transmitter power or antenna gain.

Before you start, check to see if your station is exempt from the evaluation requirement. If the transmitter power (PEP) to the antenna is less than the levels shown in **Table 7-5** on the frequencies at which you operate, then no evaluation is required! The FCC has determined that the risk of exposure from these power levels is too small to create an exposure risk. So, if you have a 25-watt VHF/UHF mobile rig or a 5-watt handheld transceiver, there's no need for an evaluation of any sort.

If you do need to do an evaluation, don't despair. It's not as complicated as it seems. The *Ham Radio License Manual* Web page lists resources that make the job a lot easier, such as on-line exposure calculators and pre-calculated tables you can use for common antennas.

The general procedure consists of several steps:

- Start with the average power from each transmitter on each band. Use the same process discussed above, starting with full PEP and then applying the various corrections for mode and use.
- If you have long coaxial feed lines, you may want to subtract feed line losses, particularly in the 30—1500 MHz frequency range.
- Then use the ARRL tables to include the effects of antenna gain and height.
- Finally, use the tables to determine the distance required from the antenna to comply with MPE limits.

The process is the same whether you do it manually using tables or on-line with a Web page calculator. Both require the same information from you about your station and use the same tables.

Don't forget to do the evaluation for each frequency band and antenna used on that band. You can save yourself some work by performing the evaluation for the highest average power, mode, and use on each antenna.

Once the evaluation is complete, compare the minimum separation with your actual installation. Chances are, you'll find no hazard exists—most stations are simply not capable of causing a health hazard.

EXPOSURE SAFETY MEASURES

What if you do find a potential hazard? What if you are just beginning to build a station and want to avoid creating a hazard? You have plenty of options as shown in **Figure 7-5**:

- Locate antennas away from where people can get close to them and away from property lines. This is always a good idea since touching an antenna radiating even low-power signals can result in an RF burn.
- Raise the antenna. This is another good idea because it usually improves your signal in distant locations, as well.
- If you have a beam antenna, avoid pointing the antenna where people are likely to be.
- Use a lower gain antenna to reduce radiated power density or reduce transmitter power. You may find that you're able to make contacts just as well with less power or gain.

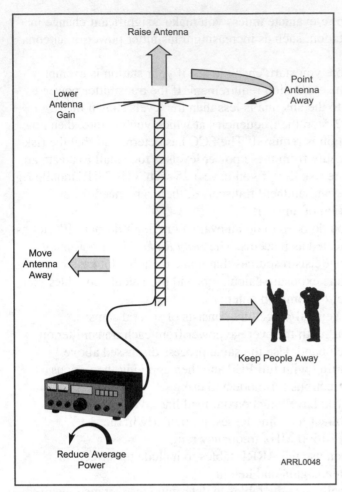

Figure 7-5 — There are many ways to reduce RF exposure to nearby people. Whatever lowers the power density in areas where people are will work. Raising the antenna will even benefit your signal strength to other stations as it lowers power density on the ground!

- Limit the average power of your transmissions by transmitting for shorter periods or even using a mode with a lower duty cycle.

Any of these techniques will reduce RF exposure to you and your neighbors. You'll likely be able to find a combination that has a minimum effect on your operations yet still makes sure you are within the MPE limits.

Even though emissions from mobile and handheld transmitters are exempt from evaluation, there are some good ways to minimize unnecessary RF exposure:

- Place mobile antennas on the roof or trunk of the car to maximize shielding of the passengers.
- Use a remote microphone to hold a handheld transceiver away from your head while transmitting.

RF exposure safety measures are easy to apply and part of good amateur practices. By understanding the reason for exposure limits and how to mitigate RF exposure hazards, you will be able to make more informed choices about designing, building, and operating your station.

Before you go on,
study questions T0C01-11.
If you have difficulty,
return to this section.

7.3 Mechanical

Just as workshop safety is important, there are plenty of mechanical aspects to Amateur Radio that generate their own safety concerns. Amateurs have been building and installing radios and antennas for 100 years, developing a large body of knowledge about the safe way to do things. Here are a few tips for jobs that you will soon be doing!

MOBILE INSTALLATIONS

Putting a radio in a car sounds pretty straightforward and it can be. However, there are a few common safety hazards that are often overlooked by a first-time installer. The most important consideration, even beyond good RF performance, is to preserve the safety of you and your passengers. Anything loose in a car can become a deadly projectile in an accident—imagine being knocked on the head by a loose radio travelling at 30 miles per hour or more! Secure *all* equipment in a vehicle, including accessories like duplexers, switches, and microphones. If possible, use *control heads* (detachable front panels) that connect to the radio with a long cable. Mount the heavier radio under a seat or in the trunk where it can't move.

Don't install the radio where it diverts your attention from the road. Don't block your vision by placing equipment on dashboards or in your field of view. (It's also a good idea to keep radios out of the direct sun.) Place the radio or control head where the controls can be easily seen without taking attention from the road for prolonged periods.

Even with your radio properly mounted you still need to be a safe driver! Follow safe operating practices by adhering to these simple rules:

- Don't operate in heavy traffic. Hang up the microphone and resume the contact later.
- Pull over to make complicated adjustments to the radio. Fumbling through a radio's menu or trying to press two buttons at once is a sure way to have an accident.

Know the traffic laws in your state about using two-way radios while driving. State legislatures have been writing laws in response to careless mobile phone use. While amateur radios are usually exempt, certain types of operating may be restricted. It may be illegal to use headphones while driving or have a speaker on too loud to hear emergency vehicles. Scanners and radio equipment that can receive public safety transmissions may be illegal in vehicles, even though amateur radios are often exempted. It is a good idea to carry your amateur license with you whenever you are operating from your car.

PUTTING UP ANTENNAS & SUPPORTS

Getting your antenna up to its lofty perch is often a little harder than it looks. The plan that looked pretty good on the ground often has to be modified once you're dealing with the actual installation. Gravity has a way of making things more difficult than you expect! Above all, follow the manufacturer's directions—they want you to have a successful experience with their product and often provide useful information in product manuals and on their Web sites. If you haven't put up an antenna before, enlist the help of a more experienced ham.

Before you can start, you should be sure your plans are in accord with any local zoning codes or by covenants or restrictions in your deed or lease. If you are putting up a very tall tower (greater than 200 feet) or an antenna near an airport, check the rules about maximum height of structures near an airport. The Federal Aviation Administration (FAA) and FCC have specific regulations about towers in these locations. These rules are covered in the *ARRL's FCC Rule Book* and are also on-line at **www.fcc.gov/wtb/rules.html**. Click on Part 17, "Construction, marking, and lighting of antenna structures."

When you're ready to put up the antenna, look carefully at the area around your antenna and any supporting structures it requires. Of course, people should not be able to

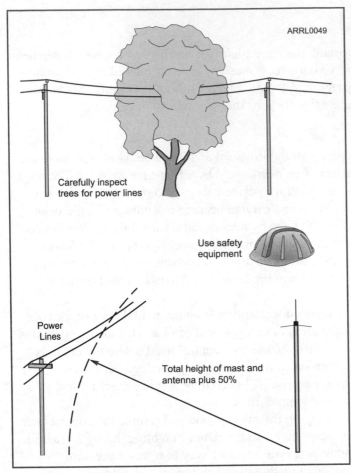

Carefully inspect
trees for power lines

ARRL0049

Use safety
equipment

Power
Lines

Total height of mast and
antenna plus 50%

Figure 7-6 — Antennas or supports falling onto a power line can result in electrocution. Take extra care in siting and erecting your antenna to avoid a deadly accident!

come in contact with the antenna accidentally. If an antenna is to be mounted at ground level, consider surrounding it with a wooden or plastic fence. If you're installing a wire antenna, make sure the feed line does not sag below head height to snag an unwary passerby. If you're in a rural area, be sure that deer or other antlered animals won't catch the feed line with their headgear.

Power lines are the enemy of antenna installers. Place all antennas and feed lines well clear of power lines, including the utility service drop to your home. **Figure 7-6** illustrates the idea. Be sure that if the antenna falls, it cannot fall onto power lines. A good rule of thumb is to separate the antenna from the nearest power line by 150% of total height of tower or mast plus antenna—a minimum of 10 feet of clearance during a fall is a must!

Trees are often used as wire supports. If you decide to throw or shoot a supporting line through or over trees be sure the projected flight path is completely safe and clear of people and power lines. A line that breaks or snags can often release a lot of energy by whipping or rebounding so wear protective gloves and goggles.

Once your antenna is in place, secure the feed line with tape or plastic wire ties. Keep all supporting guy lines above head height, if possible. Where someone can walk into them, surround the guy anchor point with a fence or flag them with colored warning or survey tape.

Grounding rules for antennas and supports must be followed according to your local electrical code. Towers should be grounded with separate 8-foot long ground rods for each tower leg, bonded to the tower and each other. A smaller antenna mast should be grounded with a heavy wire and ground rod. Guy wires must be installed according to the tower manufacturer's instructions.

Tower and Climbing Safety

It's a fact of life that antennas tend to be in high places. At VHF and UHF, the higher an antenna, the greater the line-of-sight range over which contacts can be made. Noise from sources at ground level is reduced, as well. At HF, the vertical angle at which the most power is radiated generally decreases with height above ground, so most antennas are mounted as high as is practical. Once beyond the capability of a mast or pipe support—about 40 feet or so—a steel or aluminum tower is the usual solution.

While you may not immediately decide to put up a tower yourself, it is common for amateurs to help each other with antenna projects that involve tower work. Whether you decide to work on the tower or as part of the ground crew, safety is absolutely critical. The following safety guidelines will help you safely contribute to tower projects of your own or of others. For more information about working safely on and around towers, *The ARRL Antenna Book* covers the subject in its initial chapter.

Starting with personal preparation, both climbers and ground crew should wear appropriate protective gear at all times. Each member of the crew should have their own hard hat, goggles, and heavy duty gloves suitable for working with ropes. If it's a sunny day, sun block lotion is often forgotten, but wished for after a long session outside. If you are going to climb, work boots protect the arches of your feet. Many climbers prefer footwear with a steel shank that supports the foot while standing on a narrow rung.

Before climbing or starting work, perform a thorough inspection of all equipment and installed hardware. Surprises during tower work are rarely a good thing!

• Inspect all tower guying and support hardware. Repair or tighten as necessary before anyone goes up!

• Crank-up towers must be fully nested and blocked, if necessary. Never climb a crank-up tower supported only by its support cable.

• Double-check all climbing belts and lanyards before climbing. Make sure clips and carabineers work smoothly without sticking open or closed. Replace or discard frayed straps and slings.

• Make sure all ropes and load-bearing hardware are in good condition before placing them in service.

• Double-check the latest weather report, since you don't want to get caught in a storm on the tower.

• It's a good idea to visit the bathroom before starting your climb!

Having a ground crew is important; avoid climbing alone whenever possible. If you do climb alone, take along a handheld radio. A ground crew should have enough members to do the job safely, including rendering aid if necessary. While everyone is on the ground, review the job in detail and agree on who gives instructions. If hand signals will be used, make sure everyone understands them! It also helps to rehearse the steps so that everyone knows the sequence. Does everyone know the proper knots and rope-handling technique? If not, make sure those that do are the ones that will be responsible for handling the lines.

During the job, keep distracting chatter to a minimum. One member of the crew should always be watching the climber or climbers. Stay clear of the tower base unless you need to be there because that's where dropped objects are likely to land. *Never* remove your hard hat while work is proceeding on the tower—an object dropped from 60 feet will be traveling a bit over 40 mph when it lands. Ouch!

By participating as part of a team, you'll learn the ropes, so to speak, of safe tower work. Even if you never set foot on a tower yourself, knowing how to help can make a contribution to these significant projects.

Be sure to have the proper safety equipment before beginning tower work. Also wear protective gear—hard hat, goggles, and gloves—whether you are working on the tower or as part of the ground crew.

Before you go on, try questions T0B01-09 and T0B11. If you have difficulty, return to this section.

AC hum — Unwanted 60- or 120-Hz modulation of a RF signal due to inadequate filtering in a power supply.

Adaptors — Special connectors that convert one style of connector to another.

Alternating current (ac) — Electrical current that flows first in one direction in a wire and then in the other. The applied voltage is also changing polarity. This direction reversal continues at a rate that depends on the frequency of the ac.

Amateur Satellite Corporation (AMSAT) — The organization that manages many of the amateur satellite programs.

Amateur operator — A person holding a written authorization to be the control operator of an amateur station.

Amateur Radio Emergency Service (ARES)—Sponsored by the ARRL and provides emergency communications in working with groups such as the American Red Cross and local Emergency Operations Centers.

Amateur service — A radio communication service for the purpose of self-training, intercommunication and technical investigations carried out by amateurs, that is, duly authorized persons interested in radio technique solely with a personal aim and without **pecuniary** interest.

Amateur station — A station licensed in the amateur service, including necessary equipment, used for amateur communication.

Amateur Television (ATV) — A wideband TV system that can use commercial transmission standards. ATV is only permitted on the 70-cm band (420 to 450 MHz) and higher frequencies.

American Radio Relay League (ARRL) — The national association for Amateur Radio.

Ammeter — A test instrument that measures current.

Ampere (A) — The basic unit of electrical current, also abbreviated **amps**. Current is a measure of the electron flow through a circuit. 1 Ampere is the flow of 1 **Coulomb** per second.

Amplifier — A device or piece of equipment used to increase the strength of a signal.

Amplitude modulated phone — **AM** transmission in which voice (phone) signals are used to modulate the carrier. Most AM transmission is *double-sideband* in which the signal is composed of two sidebands and a carrier. Shortwave broadcast stations use this type of AM, as do stations in the Standard Broadcast Band (535-1710 kHz). Few amateurs use double-sideband voice AM, but a variation, known as **single-sideband**, is very popular.

Amplitude modulation (AM) — The process of adding information to a signal or *carrier* by varying its amplitude characteristics.

Analog signals — A signal (usually electrical) that can have any amplitude (voltage or current) value, and whose amplitude can vary smoothly over time. Also see **digital signals**.

Antenna — A device that radiates or receives radio frequency energy.

Antenna switch — A switch used to connect one transmitter, receiver or transceiver to several different antennas.

Antenna tuner — A device that matches the antenna system input impedance to the transmitter, receiver or transceiver output impedance. Also called an *antenna-matching network, impedance matcher* or *Transmatch.*

Apogee — That point in a satellite's orbit when it is farthest from the Earth.

Automatic Position Reporting System (APRS) — A system by which amateurs can report their position automatically by radio to central servers from which their locations can be observed.

Amateur Radio Direction Finding (ARDF) — Competitions in which amateurs combine orienteering with direction finding.

Atmosphere — The mass of air surrounding the Earth. Radio signals travel through the atmosphere and different conditions in the atmosphere affect how those signal travel or propagate.

Attenuate — To reduce the strength of a signal.

Audio frequency (AF) signal — An ac electrical signal in the range of 20 hertz to 20 kilohertz (20,000 hertz). This is called an audio signal because your ears respond to sound waves in the same frequency range.

Automatic Gain Control (AGC) — Receiver circuitry used to maintain a constant audio output level.

Automatic Level Control (ALC) — Transmitter circuitry that prevents excessive modulation of an AM or SSB signal.

Automatic control — A station operating under the control of devices or procedures that insure compliance with FCC rules.

Autopatch — A device that allows repeater users to make telephone calls through a repeater.

Balun — Contraction of "balanced to unbalanced". A device to couple a balanced load to an unbalanced feed line or device, or vice versa.

Band-pass filter (BPF) — A circuit that allows signals to pass through it only if they are within a certain range of frequencies. It attenuates signals above and below this range.

Band plan — Organization of communications activity on a frequency band by general consensus.

Bandwidth — (1) Bandwidth describes the range of frequencies that a radio signal occupies. (2) FCC Part 97 defines bandwidth for regulatory purposes as "The width of a frequency band outside of which the mean power is attenuated at least 26 dB below the mean power of the transmitted signal within the band." [§97.3 (8)]

Battery — A device that converts chemical energy into electrical energy.

Battery pack — A package of several individual batteries connected together (usually in series to provide higher voltages) and treated as a single battery.

Beacon station — An amateur station transmitting communications for the purposes of observation of propagation and reception or other related experimental activities.

Beam antenna — A directional antenna. A beam antenna must be rotated to provide coverage in different directions.

Block diagram — A drawing using boxes to represent sections of a complicated device or process. The block diagram shows the connections between sections. A block diagram shows the internal functions of a complex piece of equipment without the unnecessary detail of a schematic diagram.

BNC connector — A type of connector for RF signals.

Broadcasting — Transmissions intended to be received by the general public, either direct or relayed.

Bug — A mechanical Morse key that uses a spring to send dots automatically.

Call district — The ten administrative areas established by the FCC.

Call sign — The letters and numbers that identify a specific amateur and the country in which his or her license was granted.

Calling frequency — A frequency on which amateurs establish contact before moving to a different frequency. Usually used by hams with a common interest or activity.

Capacitance — A measure of the ability of a capacitor to store energy in an *electric field*.

Capacitor — An electrical component usually formed by separating two conductive plates with an insulating material. A capacitor stores energy in an *electric field*.

CB — Citizen's Band. An unlicensed radio service operating near 27 MHz intended for use by individuals and businesses over ranges of a few miles. Also known as "11-meters" for the wavelength of its signals.

Centi (or lower case c) — The metric prefix for 10^{-2}, or divide by 100.

Certificate of Successful Completion of Examination (CSCE) — A document that verifies that an individual has passes one or more exam elements. A CSCE is good for 365 days and may be used as evidence of having passed an element at any other amateur license exam session.

Characteristic impedance — The ratio of RF voltage and current as power moves along a feed line.

Chassis ground — The common connection for all parts of a circuit that connect to the metal enclosure of the circuit. Chassis ground is usually connected to the negative side of the power supply.

Choke filter — A type of low-pass filter that blocks RF current.

Circuit breaker — A protective component that opens a circuit or *trips* when an excessive amount of current flow occurs.

Closed repeater — A repeater that restricts access to members of a certain group of amateurs.

Closed circuit — An electrical circuit with an uninterrupted path for the current to follow. Turning a switch on, for example, closes or completes the circuit, allowing current to flow. Also called a **complete circuit**.

Coaxial cable — Coax (pronounced kó-aks). A type of feed line with one conductor inside the other and both sharing a concentric central axis.

Color code — A system in which numerical values are assigned to various colors. Colored stripes representing the different values are painted on the body of resistors and sometimes other components to show their value.

Common Mode — Currents that flow equally on all conductors of a multiconductor cable, such as speaker wires or telephone cables.

Communications emergency — A situation in which communications is required for immediate safety of human life or protection of property.

Complete circuit — An electrical circuit with an uninterrupted path for the current to follow. Turning a switch on, for example, closes or completes the circuit, allowing current to flow. Also called a **closed circuit**.

Conductor — A material whose electrons move freely in response to voltage, so an electrical current can pass through it.

Continuous wave (CW) — Radio communications transmitted by on/off keying of a continuous radio-frequency signal. Another name for international Morse code.

Control operator — An amateur operator designated by the licensee of a station to be responsible for the transmissions of an amateur station.

Control point — The locations at which a station's control operator function is performed.

Controlled environment — Any area in which an RF signal may cause radiation exposure to people who are aware of the radiated electric and magnetic fields and who can exercise some control over their exposure to these fields. The FCC generally considers amateur operators and their families to be in a controlled RF exposure environment to determine the maximum permissible exposure levels.

Coulomb (C) — The basic unit of charge. 1 Coulomb is the quantity of 6.25×10^{18} electrons. 1 Ampere equals the flow of 1 Coulomb of electrons per second.

Courtesy tone — A tone or beep transmitted by a repeater to indicate that it is okay for the next station to begin transmitting. The courtesy tone is designed to allow a pause between transmissions on a repeater, so other stations can call. It also indicates that the **time-out timer** has been reset.

CQ — "Calling any station": the general call when requesting a conversation with anyone.

Crossband — Able to receive and transmit on different amateur frequency bands. For example, a repeater might retransmit at 2 meters a signal received on 70 cm.

CTCSS — Continuous Tone Coded Squelch System. A low frequency tone system used on most repeaters. When added to a carrier, a CTCSS tone allows a receiver to output the received information. Also called **PL** or **sub-audible tone**.

Current — A flow of electrons in an electrical circuit.

CW (Morse code) — Radio communications transmitted by on/off keying of a continuous radio-frequency signal. Another name for international Morse code.

D region — The lowest region of the ionosphere. The D region (or layer) contributes very little to short-wave radio propagation. It acts mainly to absorb energy from radio waves as they pass through it. This absorption has a significant effect on signals below about 7.5 MHz during daylight.

Data modes — Computer-to-computer communication, such as by **packet radio** or **radioteletype (RTTY)**, which can be used to transmit and receive computer characters, or digital information.

DE — The Morse code abbreviation for "from" or "this is."

Deceptive signals — Transmissions that are intended to mislead or confuse those who may receive the transmissions. For example, distress calls transmitted when there is no actual emergency are false or deceptive signals.

Decibel (dB) — The smallest change in sound level that can be detected by the human ear. In electronics decibels are used to express ratios of power, voltage, or current. One dB = 10 log (power ratio) or 20 log (voltage or current ratio).

Deci (or lower case d) — The metric prefix for 10^{-1}, or divide by 10.

Delta loop antenna — A variation of the quad antenna with triangular elements.

Detector — The stage in a receiver in which the modulation (voice or other information) is recovered from the RF signal.

Deviation — The change in frequency of an FM carrier due to a modulating signal.

Diffract — To alter the direction of a radio wave as it passes by or through the edges of obstructions such as buildings or hills.

Digital communications — see **data modes**.

Digital signal — (1) A signal (usually electrical) that can only have certain specific amplitude values, or steps—usually two; 0 and 1 or ON and OFF. (2) On the air, a digital signal is the same as a **data signal**.

Diode — An electronic component that allows electric current to flow in only one direction.

Dipole antenna — An antenna consisting of two symmetrical linear halves. Also see **Half-wave dipole**. A dipole need not be $1/2$ wavelength long, nor is it required to have a feed point in the middle.

Direct current (dc) — Electrical current that flows in one direction only.

Direct detection — A type of **RF interference** caused by a device being disrupted by the presence of an RF signal it is not intended to receive.

Directional wattmeter (see **Wattmeter**)

Director — A parasitic element in front of the driven element in a directional antennas.

Distress call — A transmission made in order to attract attention in an emergency. (See also **MAYDAY** and **SOS**)

Doppler shift or effect — A change in observed frequency of a signal caused by relative motion between the transmitter and receiver. Your ears hear Doppler shift when a car or train drives past you and you hear the pitch of the engine noise change. You will have to adjust your receive frequency to hear a satellite as it passes overhead because of Doppler shift.

Doubling — The undesirable act of two or more operators transmitting at the same time on the same frequency. Both operators are usually unaware of the other's presence, sometimes during the entire transmission!

Driven element — The part of an antenna that connects directly to the feed line.

Digital Signal Processing (DSP) — The process of converting an **analog signal** to **digital** form and using a microprocessor to process the signal in some way such as filtering or reducing noise.

Downlink — The frequency or frequency range on which a satellite transmits to the ground.

Dual-band antenna — An antenna designed for use on two different amateur bands.

Dummy antenna or **dummy load** — A station accessory that allows you to test or adjust transmitting equipment without sending a signal out over the air. Also called **dummy load**.

Duplex — (1) Transmitting on one frequency and receiving on another, such as for repeater operation. (2) A mode of communications (also known as *full duplex*) in which a user transmits on one frequency and receives on another frequency simultaneously. This is in contrast to half duplex, where the user transmits at one time and receives at another time.

Duplexer — A device that allows radios on two different bands to share a single antenna. Duplexers are often used to allow a dual-band radio to use a single dual-band antenna.

Duty cycle — A measure of the amount of time a transmitter is operating at full output power during a single transmission. A lower duty cycle reduces **RF radiation** exposure for the same PEP output.

DX — Distance, distant stations, foreign countries.

E region — The second lowest ionospheric region, the E region (or layer) exists only during the day. Under certain conditions, it may refract radio waves enough to return them to Earth.

Earth ground — A circuit connection to a ground rod driven into the Earth or to a metallic cold-water pipe that goes into the ground.

Earth station — An amateur station located on, or within 50 km of, the Earth's surface intended for communications with space stations or with other Earth stations by means of one or more other objects in space.

Earth-Moon-Earth (EME) or **Moonbounce** — A method of communicating with other stations by reflecting radio signals off the Moon's surface.

Echolink — A system of linking repeaters and computer-based users by using the Voice-Over-Internet Protocol.

Electric field — An electric field exists in a region of space if an electrically charged object placed in the region is subjected to an electrical force.

Electromagnetic wave — A wave of energy composed of an electric and magnetic field.

Electromotive force (EMF) — The force or pressure that pushes a current through a circuit.

Electron — A tiny, negatively charged particle, normally found in the volume surrounding the nucleus of an atom. Moving electrons make up an electrical current.

Electronic keyer — A device that makes it easier to send well-timed Morse code. It sends a continuous string of either dots or dashes, depending on which lever of the *paddle* is pressed.

Element — The conducting part or parts of an antenna designed to radiate or receive radio waves.

Elmer — A ham radio mentor or teacher.

Emergency — A situation where there is a danger to human life or property.

Emergency communications — Communications conducted under adverse conditions where normal channels of communications are not available.

Emergency traffic — Messages with life and death urgency or requests for medical help and supplies that leave an area shortly after an emergency.

Emission — The transmitted signal from an amateur station.

Emission privilege — Permission to use a particular emission type (such as Morse code or voice).

Emission types — Term for the different modes authorized for use on the Amateur Radio bands. Examples are CW, SSB, RTTY and FM.

Encoding — Changing the form of a signal into one suitable for storage or transmission. *Decoding* is the process of returning the signal to its original form.

Encryption — Changing the form of a signal into a privately-known format intended to obscure the meaning of the signal. *Decryption* is the process of reversing the encoding.

Energy — The ability to do work; the ability to exert a force to move some object.

Extended-coverage receiver — A receiver that tunes frequencies from around 30 MHz to several hundred MHz or into the GHz frequencies. Also known as a **wide-range receiver**.

F region — A combination of the two highest ionospheric regions (or layers), the F1 and F2 regions. The F region refracts radio waves and returns them to Earth. Its height varies greatly depending on the time of day, season of the year and amount of sunspot activity.

False or deceptive signals — Transmissions that are intended to mislead or confuse those who may receive the transmissions. For example, distress calls transmitted when there is no actual emergency are false or deceptive signals.

Farad (F) — The basic unit of capacitance.

Federal Communications Commission (FCC) — Federal agency in the United States that regulates use and allocation of the frequency spectrum among many different services, including Amateur Radio.

Federal Registration Number (FRN) — An identification number assigned to an individual by the FCC to use when performing license modification or renewal.

Feed line — The wires or cable used to connect a transmitter, receiver or transceiver to an antenna. The feed line connects to an antenna at its feed point. Also see **Transmission line**.

Feed point — The point at which a feed line is electrically connected to an antenna.

Feed point impedance — The ratio of RF voltage to current at the feed point of an antenna.

Ferrite — A ceramic material that absorbs RF energy. Ferrite is often formed into beads or cores so that it may be placed on cables to prevent RF signal from flowing along the cable's outer conductor.

Filter — A circuit that will allow some signals to pass through it but will greatly reduce the strength of others.

Fixed resistor — An electronic component specifically designed to oppose or control current through a circuit. The **resistance** value of a fixed resistor cannot be changed or adjusted.

Form 605 — An FCC form that serves as the application for your Amateur Radio license, or for modifications to an existing license.

Forward power — The power traveling from the transmitter to the antenna along a transmission line.

Fox hunting — Exercises in which a hidden transmitter (the fox) is located in order to test direction-finding skills. Also called a *bunny hunt*.

Frequency — The number of complete cycles of an alternating current that occur per second.

Frequency band — A continuous range of frequencies in which one type of communications is authorized. An **amateur band** is a frequency band in which amateur communications take place.

Frequency coordination — Allocating repeater input and output frequencies to minimize interference between repeaters and to other users of the band.

Frequency coordinator — An individual or group that recommends repeater frequencies to reduce or eliminate interference between repeaters operating on or near the same frequency in the same geographical area.

Frequency discriminator — A type of detector used in some FM receivers.

Frequency modulated phone — The type of signals used to communicate by voice (phone) over most repeaters. FM broadcast stations and most professional communications (police, fire, taxi) use FM. VHF/UHF FM phone is the most popular amateur mode.

Frequency modulation (FM) — The process of adding information to an RF signal or *carrier* by varying its frequency characteristics.

Frequency privilege — Permission to use a particular group of frequencies.

Front-end overload — Interference to a receiver caused by a strong signal that causes the receiver's sensitive input circuitry ("front end") to be overloaded or saturated. Front-end overload results in distortion of the desired signal and the generation of unwanted spurious signals within the receiver. See also **receiver overload**.

FRS — Family Radio Service. An unlicensed radio service that uses low-power radios operating near 460 MHz and intended for short-range communications by family members.

Fuse — A thin metal strip mounted in a holder. When too much current passes through the fuse, the metal strip melts and opens the circuit.

Gain — (1) Focusing of an antenna's radiated energy in one direction. Gain in one direction means that gain in other directions is diminished. (2) The amount of amplification of a signal in a piece of equipment, such as AF Gain (volume) or RF Gain (sensitivity).

General-coverage receiver — A receiver used to listen to a wide range of frequencies, not just specific bands. Most general-coverage receivers tune from frequencies below the AM broadcast band (550 - 1700 kHz) to around 30 MHz. (See also **extended-coverage receiver**.)

GFI — Ground-fault interrupting circuit breaker that opens a circuit when an imbalance of current flow is detected between the hot and neutral wires of an ac power circuit.

Giga (or lower case G) — The metric prefix for 10^9, or multiply by 1,000,000,000.

GMRS — General Mobile Radio Service. A licensed radio service operating 460 MHz intended for family businesses and members to communicate within a city or region.

Go kit — A pre-packaged collection of equipment or supplies kept at hand to allow an operator to quickly report where needed in time of need.

Grace period — The time FCC allows following the expiration of an amateur license to renew that license without having to retake an examination. Those who hold an expired license may not operate an amateur station until the license is reinstated.

Grid square — A locator in the Maidenhead Locator System.

Ground connection — A connection made to the earth for electrical safety. This connection can be made inside (to a metal cold-water pipe) or outside (to a **ground rod**).

Ground rod — A copper or copper-clad steel rod that is driven into the earth. A heavy copper wire or strap connects all station equipment to the ground rod.

Ground-plane — A conducting surface of continuous metal or discrete wires that acts to create an electrical image of an antenna. **Ground-plane antennas** require a ground-plane in order to operate properly.

Ground-wave propagation — The method by which radio waves travel along the Earth's surface.

Ham-band receiver — A receiver designed to receive only frequencies in the amateur bands.

Half-wave dipole — A basic antenna used by radio amateurs. It consists of a length of wire or tubing with a feed point at the center. The entire antenna is ½ wavelength long at the desired operating frequency.

Hand-held radio — A VHF or UHF transceiver that can be carried in the hand or pocket.

Harmful interference — Interference that seriously degrades, obstructs or repeatedly interrupts a radio communication service operating in accordance with the Radio Regulations. [§97.3 (a) (22)]

Harmonics — Signals from a transmitter or oscillator occurring on whole-number multiples (2×, 3×, 4×, etc) of the original or *fundamental* frequency.

Health and Welfare traffic – Messages about the well-being of individuals in a disaster area. Such messages must wait for **Emergency** and **Priority traffic** to clear, and results in advisories to those outside the disaster area awaiting news from family and friends.

Henry (H) — The basic unit of inductance.

Hertz (Hz) — An alternating-current frequency of one cycle per second. The basic unit of frequency.

High frequency (HF) — The term used for the frequency range between 3 MHz and 30 MHz. The Amateur HF bands are where you are most likely to make long-distance (worldwide) contacts.

High-pass filter (HPF) — A filter designed to pass signals above a specified *cutoff* frequency, while attenuating lower-frequency signals.

Impedance — The opposition to electric current in a circuit. Impedance includes both reactance and resistance, and applies to both alternating and direct currents.

Impedance match — To adjust impedances to be equal or the case in which two impedances are equal. Usually refers to the point at which a feed line is connected to an antenna or to transmitting equipment. If the impedances are different, that is a *mismatch*.

Impedance-matcher — A device that matches one impedance level to another. For example, it may match the impedance of an antenna system to the impedance of a transmitter or receiver. Amateurs also call such devices a Transmatch, antenna-matcher or **antenna tuner**.

Inductance — A measure of the ability of a coil to store energy in a *magnetic field*.

Inductor — An electrical component usually composed of a coil of wire wound on a central core. An inductor stores energy in a *magnetic field*.

Input frequency — A repeater's receiving frequency. To use a repeater, transmit on the input frequency and receive on the **output frequency**.

Insulator — A material whose electrons do not move easily, so that an electric current cannot pass through it (within voltage limits).

Integrated circuit (IC) — A compound electronic component composed of many individual components in a single package.

Intermediate frequency (IF) — The stages in a receiver that follow the input amplifier and mixer circuits. Most of the receiver's gain and selectivity are achieved at the IF stages.

International Telecommunications Union (ITU) — The organization of the United Nations responsible for coordinating international telecommunications agreements.

Internet Repeater Linking Project (IRLP) — A system of linking repeaters by using the Voice-Over-Internet Protocol.

Ionizing radiation — Electromagnetic radiation that has sufficient energy to knock electrons free from their atoms, producing positive and negative ions. X-rays, gamma rays and ultraviolet radiation are examples of ionizing radiation.

Ionosphere — A region of electrically charged (ionized) gases high in the atmosphere. The ionosphere bends radio waves as they travel through it, returning them to Earth. Also see **sky-wave propagation**.

Isotropic antenna — An antenna that radiates and receives equally in all possible directions.

K — The Morse code abbreviation for "end of transmission" or "go ahead." Any station may respond.

Keplerian elements — Mathematical values for a satellite's orbit that can be used to compute the position of a satellite at any point in time, for any position on Earth.

Key — A manually operated switch that turns a transmitter on and off to send Morse code.

Keyer or **electronic keyer** — A piece of equipment that generates Morse code automatically.

Kilo (or lower case k) — The metric prefix for 10^3, or multiply by 1000.

Lightning protection — Methods to prevent lightning damage to your equipment (and your house), such as unplugging equipment, disconnecting antenna feed lines and using a lightning arrestor.

Line-of-sight propagation — The term used to describe VHF and UHF propagation in a straight line directly from one station to another.

Linear amplifier — Also known as a **linear**, a piece of equipment that amplifies the output of a transmitter, often to the full legal amateur power limit of 1500 W PEP.

Local control — Operation of a station with a control operator physically present at the transmitter.

Log — The documents or log of a station that detail operation of the station. They can be used as supporting evidence, and for troubleshooting interference-related problems or complaints.

Loop antenna — An antenna with element(s) constructed as continuous lengths of wire or tubing.

Lower sideband (LSB) — (1) In an AM signal, the sideband located below the carrier frequency. (2) The common single-sideband operating mode on the 40, 80 and 160-meter amateur bands.

Low-pass filter (LPF) — A filter designed to pass signals below a specified *cutoff* frequency, while attenuating higher-frequency signals.

Malicious (willful) interference — Intentional, deliberate obstruction of radio transmissions.

Maximum useable frequency (MUF) — The highest-frequency radio signal that will reach a particular destination using **sky-wave propagation**, or *skip*. The MUF may vary for radio signals sent to different destinations.

MAYDAY — From the French *m'aidez* (help me), MAYDAY is used when calling for emergency assistance in voice modes.

Maximum Permissible Exposure (MPE) — The maximum intensity of RF radiation to which a human being may be exposed. FCC Rules establish maximum permissible exposure values for humans to RF radiation. [§1.1310 and §97.13 (c)]

Mega (or capital M) — The metric prefix for 10^6, or times 1,000,000.

Memory channel — Frequency and mode information stored by a radio and referenced by an alphanumeric designator.

Meteor Scatter — Communicating by reflecting signals off of the ionized trails left by meteors in the upper atmosphere.

Metric prefixes — A series of terms used in the metric system of measurement. We use metric prefixes to describe a quantity as compared to a basic unit. The metric prefixes indicate multiples of 10.

Metric system — A system of measurement developed by scientists and used in most countries of the world. This system uses a set of prefixes that are multiples of 10 to indicate quantities larger or smaller than the basic unit.

Micro (or μ) — The metric prefix for 10^{-6}, or divide by 1,000,000.

Microphone — A device that converts sound waves into electrical energy. (abbreviated **MIC** or **MIKE**)

Microwave — Radio waves or signals with frequencies greater than 1000 MHz (1 GHz). This is not a strict definition, just a conventional way of referring to those frequencies.

Milli (or lower case m) — The metric prefix for 10^{-3}, or divide by 1000.

Mixer — Circuitry that combines two signals and generates signals at both their sum and difference frequencies. Mixers are used in receivers and transmitters to convert signals from one frequency to another.

Mobile station — A radio transmitter designed to be mounted in a vehicle. A push-to-talk (PTT) switch generally activates the transmitter. Any station that can be operated on the move, typically in a car, but also on a boat, a motorcycle, truck or RV.

Mode — The combination of a type of information and a method of transmission. For example, FM radiotelephony or *FM phone* consists of using FM modulation to carry voice information.

Modem — Short for *mo*dulator/*dem*odulator. A modem changes data into audio signals that can be transmitted by radio and demodulates a received signal to recover transmitted data.

Modulate or modulation — The process of adding information to an RF signal or *carrier* by varying its amplitude, frequency, or phase.

Morse code (see **CW**).

Multiband antenna — An antenna capable of operating on more than one amateur frequency band, usually using a single feed line.

Multihop propagation — Long-distance radio propagation using several skips or hops between the Earth and the ionosphere.

Multimeter — An electronic test instrument used to measure current, voltage and resistance in a circuit. Describes all meters capable of making these measurements, such as the volt-ohm-milliammeter (VOM), vacuum-tube voltmeter (VTVM) and field-effect transistor VOM (FET VOM).

Multimode radio — Transceiver capable of SSB, CW and FM operation.

Multiple Protocol Controller (MPC) — A piece of equipment that can act as a **TNC** for several **protocols**.

N or **type N connector** — A type of RF connector.

National Electrical Code — A set of guidelines governing electrical safety, including antennas.

Net — An formal system of operation in order to exchange or manage information

Net control station (NCS) — The station in charge of a net.

Network — A term used to describe several digital stations linked together to relay data over long distances.

National Incident Management System (NIMS) — The method by which emergency situations are managed by US public safety agencies.

Noise blanker — A circuit that mutes the receiver during noise pulses.

Noise reduction — Removing random noise from a receiver's audio output.

Nonionizing radiation — Electromagnetic radiation that does not have sufficient energy to knock electrons free from their atoms. Radio frequency (RF) radiation is nonionizing.

Notch filter — A filter that removes a very narrow range of frequencies, usually from a receiver's audio output to remove continuous tones.

Offset frequency — The difference between a repeater's transmitter and receiver frequencies. Also known as the *repeater split*.

Ohm — The basic unit of electrical resistance.

Ohm's Law — A basic law of electronics. Ohm's Law states the relationship between voltage (E), current (I) and resistance (R). The voltage applied to a circuit is equal to the current through the circuit times the resistance of the circuit (E = IR).

Ohmmeter — A device used to measure resistance.

Omnidirectional — An antenna that radiates and receives equally in all horizontal directions.

One-way communications — Radio signals not directed to a specific amateur radio station, or for which no reply is expected. The FCC Rules provide for limited types of one-way communications on the amateur bands. [§97.111 (b)]

Open circuit — An electrical circuit that does not have a complete path, so current can't flow through the circuit.

Open repeater — A repeater that can be used by all hams who have a license that authorizes operation on the repeater frequencies.

Operator/primary station license — An amateur license actually includes two licenses in one. The operator license is that portion of an Amateur Radio license that gives permission to operate an amateur station. The **primary station license** is that portion of an Amateur Radio license that authorizes an amateur station at a specific location. The station license also lists the call sign of that station.

Oscillate — To vibrate continuously at a single frequency. An **oscillator** is a device or circuit that generates a signal at a single frequency.

Output frequency — A repeater's transmitting frequency. To use a repeater, transmit on the **input frequency** and receive on the output frequency.

Packet radio — A system of digital communication whereby information is broken into short bursts. The bursts ("packets") also contain addressing and error-detection information.

Paddle — Similar to a **key** or **bug**, a paddle has a pair of contacts operated by one or two levers that is used to control an electronic **keyer** that generates Morse code automatically.

Parallel circuit — An electrical circuit in which the electrons may follow more than one path in traveling between the negative supply terminal and positive terminal.

Parallel-conductor line — A type of transmission line that uses two parallel wires spaced apart from each other by insulating material. Also known as *open-wire, ladder, or window line*.

Parasitic element — Part of a directional antenna that derives energy from mutual coupling with the driven element. Parasitic elements are not connected directly to the feed line.

Part 15 — The section of the FCC's rules that deal with unlicensed devices likely to transmit or receive RF signals.

Part 97 — The section of the FCC's rules that regulate Amateur Radio.

Peak envelope power (PEP) — The average power of an RF signal at its largest amplitude peak.

Pecuniary — Payment of any type, whether money or other goods. Amateurs may not operate their stations in return for any type of payment.

Perigee — That point in the orbit of a satellite when it is closest to the Earth.

Phase — A measure of position in time within a repeating waveform, such as a sine wave. Phase is measured in degrees or radians. There are 360 degrees or 2π radians in one complete cycle.

Phase modulation (PM) — The process of adding information to a signal by varying its phase characteristics. Phase modulation is very similar to **FM** and PM signals can be received by FM receivers.

Phone — Another name for voice communications.

Phone emission — The FCC name for voice or other sound transmissions.

Phonetic alphabet — Standard words used on voice modes to make it easier to understand letters of the alphabet, such as those in call signs. The call sign KA6LMN stated phonetically is *Kilo Alfa Six Lima Mike November*.

Pico (or lower case p) — The metric prefix for 10^{-12}, or divide by 1,000,000,000,000.

PL (see CTCSS) — Private Line. PL is a Motorola trademark.

Polarization — The orientation of the electrical-field of a radio wave. An antenna that is parallel to the surface of the earth, such as a dipole, produces horizontally polarized waves. One that is perpendicular to the earth's surface, such as a quarter-wave vertical, produces vertically polarized waves. An antenna that has both horizontal and vertical polarization is said to be circularly polarized.

Portable device — Generally considered to be a radio transmitting device designed to be transported easily and set up for operation independently of normal infrastructure. For purposes of RF exposure regulations, a portable device is one designed to have a transmitting antenna that is generally within 20 centimeters of a human body.

Potentiometer — Another name for a **variable resistor**. The resistance value of a potentiometer can be changed over a range of values without removing it from a circuit.

Power — The rate of energy consumption or expenditure. We calculate power in an electrical circuit by multiplying the voltage applied to the circuit times the current through the circuit (P = IE).

Power amplifier — See **linear amplifier**

Power supply — A circuit that provides a direct-current output at some desired voltage from an ac input voltage.

Preamplifier — An amplifier placed ahead of a receiver's input circuitry to increase the strength of a received signal. Preamplifier circuits are often included in a receiver and may be turned on or off. Preamplifiers for VHF, UHF, and microwave frequencies are sometimes located at the antenna to amplify signals before loss in the feed line reduces their strength.

Prefix — The leading letters and numbers of a call sign that indicate the country in which the call sign was assigned.

Primary service — When a frequency band is shared among two or more different radio services, the primary service is preferred. Stations in the **secondary service** must not cause harmful interference to, and must accept interference from stations in the primary service. [§97.303]

Primary station license — An amateur license actually includes two licenses in one. The **operator license** is that portion of an Amateur Radio license that gives permission to operate an amateur station. The primary station license is that portion of an Amateur Radio license that authorizes an amateur station at a specific location. The station license also lists the call sign of that station.

Priority traffic — Emergency-related messages, but not as important as **Emergency traffic**.

Procedural signals (prosign) — For Morse code communications, one or two letters sent as a single character. Amateurs use prosigns in CW contacts as a short way to indicate the operator's intention. Some examples are к for "Go Ahead or AR for "End of Message." (A bar over the letters is used to indicate that the prosign is sent as one character.) For phone communications, words such as "Break" or "Over" that control the flow of the communications.

Product detector — A type of mixer circuit that allows a receiver to demodulate CW and SSB signals.

Propagation — The process through which radio waves travel.

Protocol — A method of encoding, packaging, exchanging, and decoding digital data.

Push to talk (PTT) — Turning a transmitter on and off manually with a switch, usually thumb- or foot-activated.

Q signals — Three-letter symbols beginning with Q used on CW to save time and to improve communication. Some examples are QRS (send slower), QTH (location), QSO (ham conversation) and QSL (acknowledgment of receipt).

Q system — A method of providing signal quality reports on a scale of 1 ("Q1") to 5 ("Q5").

QSL card — A postcard that serves as a confirmation of communication between two hams. QSL is a Q-signal meaning "received and understood."

QSO — A conversation between two radio amateurs. QSO is a Q-signal meaning "I am in contact."

Quad antenna — An antenna built with its elements in the shape of four-sided loops.

Quarter-wavelength vertical antenna — An antenna constructed of a quarter-wavelength long radiating element placed perpendicular to the earth.

Radiation pattern — A graph showing how an antenna radiates and receives in different directions. An *azimuthal pattern* shows radiation in horizontal directions. An *elevation pattern* shows how an antenna radiates and receives at different vertical angles.

Radio Amateur Civil Emergency Service (RACES) — A part of the Amateur Service that provides radio communications for civil defense organizations during local, regional or national civil emergencies.

Radio frequency (RF) exposure — FCC Rules establish maximum permissible exposure (MPE) values for humans to RF radiation. [§1.1310 and §97.13 (c)]

Radio frequency (RF) radiation or **waves** — Electromagnetic energy that travels through space without wires.

Radio frequency (RF) signals — RF signals are generally considered to be any electrical signals with a frequency higher than 20,000 Hz, up to 300 GHz.

Radio-frequency interference (RFI) — Disturbance to electronic equipment caused by radio-frequency signals.

Radiogram — A formal message exchanged via radio.

Radio horizon — The most distant point to which radio signals can be sent directly without reflections.

Radioteletype (RTTY) — Radio signals sent from one teleprinter machine to another machine. Anything that one operator types on his teleprinter will be printed on the other machine. Also known as narrow-band direct-printing telegraphy.

Ragchew — An informal conversation.

Range — The longest distance over which radio signals can be exchanged.

Receiver — A device that converts radio waves into signals we can hear or see. (abbreviated **RCVR**)

Receiver overload — Interference to a receiver caused by a RF signal too strong for the receiver input circuits. A signal that overloads the receiver RF amplifier (front end) causes front-end overload. Receiver overload is sometimes called *RF overload.*

Reciprocal operating authority — Permission for amateur radio operators from another country to operate in the US using their home license. This permission is based on various treaties between the US government and the governments of other countries.

Rectifier — A diode intended for use with high current or voltage in power supplies.

Reflected power — The power that returns to the transmitter from the antenna along a transmission line.

Reflection — Signals that travel by **line-of-sight propagation** are reflected by large objects like buildings.

Reflector — A parasitic element behind the driven element in a directional antennas.

Refract — Bending of an electromagnetic wave as it travels through materials with different properties. Light refracts as it travels from air into water. Radio waves refract as they travel through the ionosphere. If the radio waves refract enough they will return to Earth. This is the basis for long-distance communication on the **HF bands**.

Region — One of the three administrative areas defined by the **ITU**.

Remote control — Operation of a station in which the control functions of the station are operated by a control operator over a control link.

Repeater station — A station that retransmits the signals of other stations to give them greater range.

Resistance — The ability to oppose an electric current.

Resistor — An electronic component specifically designed to oppose or control current through a circuit.

Resonance — The condition of an applied signal or wave having the same frequency as the resonant frequency of a tuned circuit or antenna.

Resonant frequency — The desired operating frequency of a tuned circuit. In an antenna, the resonant frequency is one where the feed-point impedance is composed only of resistance.

RF burn — A burn produced by coming in contact with exposed RF voltages.

RF carrier — A steady radio frequency signal that is modulated to add an information signal to be transmitted. For example, a voice signal is added to the RF carrier to produce a **phone emission** signal.

RF feedback — Distortion caused by RF signals disturbing the function of an audio circuit.

RF overload — Another term for receiver overload.

RF safety — Preventing injury or illness to humans from the effects of radio-frequency energy.

Rig — The radio amateur's term for a transmitter, receiver or transceiver.

RST — A system of numbers used for signal reports: R is readability, S is strength and T is tone. (On phone, only R and S reports are used.)

Rubber duck antenna — A flexible rubber-coated antenna that is inexpensive, small, lightweight and difficult to break. Rubber ducks are used mainly with hand-held VHF or UHF transceivers.

S meter — A meter that provides an indication of the relative strength of received signals.

Safety interlock — A switch that automatically turns off power to a piece of equipment when the enclosure is opened.

Scattering — Radio wave propagation by means of multiple reflections in the layers of the atmosphere or from an obstruction.

Schematic diagram — A drawing that describes the electrical connections in a piece of electric or electronic equipment.

Schematic symbol — A standardized symbol used to represent an electrical or electronic circuit component on a schematic diagram.

Secondary service — When a frequency band is shared among two or more different radio services, the **primary service** is preferred. Stations in the secondary service must not cause harmful interference to, and must accept interference from stations in the primary service. [§97.303]

Selectivity — The ability of a receiver to distinguish between signals. Selectivity is important when many signals are present and when it is desired to receive weak signals in the presence of strong signals.

Sensitivity — The ability of a receiver to detect weak signals.

Series circuit — An electrical circuit in which all the electrons must flow through every part of the circuit because there is only one path for the electrons to follow.

Shack — The room where an Amateur Radio operator keeps his or her station equipment.

Shielding — Surrounding an electronic circuit to block RF signals from being radiated or received.

Short circuit — An electrical circuit in which the current does not take the desired path, but finds a shortcut instead. Often the current flows directly between the negative power-supply terminal and the positive one, bypassing the rest of the circuit.

Sidebands — The sum or difference frequencies generated when an RF carrier is mixed with an audio signal. Single-sideband phone (SSB) signals have an upper sideband (USB — that part of the signal above the carrier) and a lower sideband (LSB — the part of the signal below the carrier). SSB transceivers allow operation on either USB or LSB.

Signal generator — A device that produces a low-level signal that can be set to a desired frequency.

Signal report — An evaluation of the transmitting station's signal and reception quality.

Simplex operation — Receiving and transmitting on the same frequency.

Sine Wave — A waveform whose amplitude is equal to the sine of frequency × time.

Single sideband (SSB) phone — SSB is a form of double-sideband amplitude modulation in which one sideband and the carrier are removed. SSB is a common mode of voice operation on the amateur bands.

Skip — Propagation by means of ionospheric reflection. Traversing the distance to the ionosphere and back to the ground is called a *hop*.

Skip zone — An area of poor radio communication, too distant for ground waves and too close for sky waves.

Sky-wave propagation — The method by which radio waves travel through the ionosphere and back to Earth. Sometimes called *skip*, sky-wave propagation has a far greater range than **line-of-sight** and **ground-wave propagation**.

Slow-Scan Television (SSTV) — A television system used by amateurs to transmit pictures within a signal bandwidth allowed on the HF or VHF/UHF bands by the FCC. It takes approximately 8 seconds to send a single black and white SSTV frame, and between 12 seconds and 4$^1/_2$ minutes for the various color systems currently in use on the HF bands.

SMA connector — A type of RF connector.

SOS — A Morse code call for emergency assistance.

Space station — An amateur station located more than 50 km above the Earth's surface.

Specific absorption rate (SAR) — A term that describes the rate at which RF energy is absorbed into the human body. Maximum permissible exposure (MPE) limits are based on whole-body SAR values.

Speech compression or **processing** — Increasing the average power of a voice signal by amplifying low-level components of the signal more than high-level components.

Splatter — A type of interference to stations on nearby frequencies. Splatter occurs when a transmitter is overmodulated.

Sporadic E — A form of enhanced radio-wave propagation that occurs when radio signals are reflected from small, dense ionization patches in the E region of the ionosphere. Sporadic E is observed on the 15, 10, 6 and 2-meter bands, and occasionally on the 1.25-meter band.

Spurious emissions — Signals from a transmitter on frequencies other than the operating frequency.

Squelch — Circuitry that mutes an FM receiver when no signal is received.

SSB — Abbreviation for the **single sideband phone** mode of communication. This is the most widely used mode for phone operation on the HF bands.

Standard frequency offset — The standard transmitter/receiver frequency offset used by a repeater on a particular amateur band. For example, the standard offset on 2 meters is 600 kHz. Also see **Offset frequency**.

Standing-wave ratio (SWR) — Sometimes called voltage standing-wave ratio (VSWR). A measure of the impedance match between the feed line's characteristic impedance and the load (usually an antenna). Also, with a Transmatch in use, a measure of the match between the feed line from the transmitter and the antenna system. The system includes the Transmatch and the line to the antenna. VSWR is the ratio of maximum voltage to minimum voltage along the feed line, also the ratio of antenna impedance to feed-line impedance.

Station grounding — The practice of connecting all station equipment to a good earth ground to improve both safety and station performance.

Station license — An amateur license actually includes two licenses in one. The **operator license** is that portion of an Amateur Radio license that gives permission to operate an amateur station. The primary station license is that portion of an Amateur Radio license that authorizes an amateur station at a specific location. The station license also lists the call sign of that station.

Station records/station log — The documents or log of a station that detail operation of the station. The log can be used as supporting evidence, and for troubleshooting interference-related problems or complaints.

Stratosphere — The part of the Earth's atmosphere that extends from about 7 miles to 30 miles above the earth. Clouds rarely form in the stratosphere.

Sub-audible tone — see CTCSS

Suffix — The letters that follow a call sign prefix identifying a specific amateur.

Sunspot cycle — The number of **sunspots** increases and decreases in a predictable cycle that lasts about 11 years.

Sunspots — Dark spots on the surface of the sun. When there are few sunspots, long-distance radio propagation is poor on the higher-frequency bands. When there are many sunspots, long-distance HF propagation improves.

Surge protector — A device that limits voltage by changing from an insulator to a conductor when excessive voltage occurs. Surge protectors are used to prevent temporary or *transient* excessive voltages from damaging sensitive electronic equipment.

Switch — A device used to connect or disconnect electrical contacts.

SWR meter — A measuring instrument that can indicate when an antenna system is working well. A device used to measure SWR.

Tactical call signs — Names used to identify a location or function during local emergency communications.

Tactical communications — A first-response communications under emergency conditions that involves a few people in a small area.

Telecommand operation — A one-way radio transmission to start, change or end functions of a device at a distance.

Telegraph key — A telegraph key (also called a *straight key*) is the simplest type of Morse code sending device.

Teleprinter — A machine that can convert keystrokes (typing) into electrical impulses. The teleprinter can also convert the proper electrical impulses back into text. Computers have largely replaced teleprinters for amateur radioteletype work.

Television interference (TVI) — Interruption of television reception caused by another signal.

Temperature inversion — A condition in the atmosphere in which a region of cool air is trapped beneath warmer air.

Temporary state of communications emergency — When a disaster disrupts normal communications in a particular area, the FCC can declare this type of emergency. Certain rules may apply for the duration of the emergency.

Terminal — An inexpensive piece of equipment that can be used in place of a computer in a packet radio station.

Third-party — An unlicensed person on whose behalf communications is passed by amateur radio.

Third-party communications — Messages passed from one amateur to another on behalf of a third person.

Third-party communications agreement — An official understanding between the United States and another country that allows amateurs in both countries to participate in third-party communications.

Third-party participation — An unlicensed person participating in amateur communications. A control operator must ensure compliance with FCC rules.

Ticket — A common name for an Amateur Radio license.

Time-out timer — A device that limits the amount of time any one person can talk through a repeater.

Terminal Node Controller (TNC) — A device that acts as an interface between a computer and a radio. It includes a **modem** and implements the rules of a **protocol**.

T-R switch — Transmit-Receive switch. A circuit or device that switches an antenna between transmitter and receiver circuits or equipment.

Traffic — Formal messages exchanged via radio. *Traffic handling* is the process of exchanging traffic. A *traffic net* is a net specially created and managed to handle traffic.

Transceiver — A radio transmitter and receiver combined in one unit. (abbreviated **XCVR**)

Transistor — A solid-state device made of three layers of semiconductor material. A transistor can be used as a switch or amplifier.

Transmission line — The wires or cable used to connect a transmitter or receiver to an antenna. Also called **feed line**.

Transmitter — A device that produces radio-frequency signals. (abbreviated **XMTR**)

Trip — Activate based on some physical action (current, signal level, voltage) exceeding a threshold. A circuit breaker trips, opening a circuit, when excessive current flow occurs, for example.

Troposphere — The region in Earth's atmosphere just above the Earth's surface and below the ionosphere.

Tropospheric bending — When radio waves are bent in the troposphere, they return to Earth farther away than the visible horizon.

Tropospheric ducting — A type of VHF propagation that can occur when warm air overruns cold air (a temperature inversion).

UHF connector — A type of RF connector.

Ultra high frequency (UHF) — The term used for the frequency range between 300 MHz and 3000 MHz (3 GHz). Technician licensees have full privileges on all Amateur UHF bands.

Ultraviolet (UV) — Electromagnetic waves with frequencies higher than visible light. Literally, "above violet," which is the high-frequency end of the visible range.

Unbalanced line — Feed line with one conductor at ground potential, such as coaxial cable.

Uncontrolled environment — Any area in which an RF signal may cause radiation exposure to people who may not be aware of the radiated electric and magnetic fields. The FCC generally considers members of the general public and an amateur's neighbors to be in an uncontrolled **RF radiation** exposure environment to determine the maximum permissible exposure levels.

Unidentified communications or signals — Signals or radio communications in which the transmitting station's call sign is not transmitted.

Universal Licensing System (ULS) — FCC database for all FCC radio services.

Uplink — The frequency or frequency range on which signals are transmitted from the ground to a satellite.

Upper sideband (USB) — (1) In an AM signal, the sideband located above the carrier frequency. (2) The common single-sideband operating mode on the 20, 17, 15, 12 and 10-meter HF amateur bands, and all the VHF and UHF bands.

Vanity call — A call sign selected by an amateur instead of sequentially assigned by the FCC.

Variable resistor — A resistor whose value can be adjusted over a certain range, without removing it from a circuit.

Variable-frequency oscillator (VFO) — An oscillator used in receivers and transmitters. The frequency is set by a tuned circuit using capacitors and inductors and can be changed by adjusting the components of the tuned circuit.

Vertical antenna — A common amateur antenna whose radiating element is vertical. There are usually four or more radial elements parallel to or on the ground.

Very high frequency (VHF) — The term used for the frequency range between 30 MHz and 300 MHz. Technician licensees have full privileges on all Amateur VHF bands.

Visible horizon — The most distant point one can see by line of sight.

Voice — Any of the several methods used by amateurs to transmit speech.

Voice communications — Hams can use several voice modes, including FM and SSB.

Volt (V) — The basic unit of electrical potential or EMF.

Voltage — The EMF or electrical potential difference that causes electrons to move through an electrical circuit.

Voltmeter — A test instrument used to measure voltage.

Volunteer Examiner (VE) — A licensed amateur who is accredited by a Volunteer Examiner Coordinator (VEC) to administer amateur license examinations.

Volunteer Examiner Coordinator (VEC) — An organization that has entered into an agreement with the FCC to coordinate amateur license examinations.

Voice-Operated Transmission (VOX) — Turning a transmitter on and off under control of the operator's voice.

Watt (W) — The unit of power in the metric system. The watt describes how fast a circuit uses electrical energy.

Wattmeter — Also called a *power meter*, a test instrument used to measure the power output (in watts) of a transmitter. A directional wattmeter measures both forward and reflected power in a feed line.

Wavelength — Often abbreviated λ. The distance a radio wave travels in one RF cycle. The wavelength relates to frequency. Higher frequencies have shorter wavelengths.

Weak-signal modes — Usually SSB or CW modes, used in relation to operating on the VHF and UHF bands, where many amateurs only operate FM phone.

Whip antenna — An antenna with an element made of a single, flexible rod or tube.

Winlink — A system of email transmission and distribution using Amateur Radio for the connection between individual amateurs and mailbox stations known as *Participating Mailbox Operators (PMBO)*.

Wiring diagram — A pictorial or descriptive drawing that shows how the wiring of a piece of electric or electronic equipment is to be done.

WWV/WWVH — Radio stations run by the US NIST (National Institute of Standards and Technology) to provide accurate time and frequencies.

XCVR — Transceiver

XMTR — Transmitter

Yagi antenna — The most popular type of directional (beam) antenna. It has one driven element and one or more additional parasitic elements.

73 — Ham lingo for "best regards." Used on both phone and CW toward the end of a contact.

88 — Ham lingo for "love and kisses" (not meant literally) when concluding a contact with a female operator.

Technician Class Question Pool
Effective July 1, 2006

SUBELEMENT T1
FCC Rules, station license responsibilities
4 exam questions – 4 Groups

T1A Basis and purpose of the Amateur Radio Service, penalties for unlicensed operation, other penalties, examinations – 1 exam question

T1A01
Who is an amateur operator as defined in Part 97?
A. A person named in an amateur operator/primary license grant in the FCC ULS database
B. A person who has passed a written license examination
C. The person named on the FCC Form 605 Application
D. A person holding a Restricted Operating Permit

T1A01
(A)
[97.3(a)(1)]
p 5-3

T1A02
What is one of the basic purposes of the Amateur Radio Service as defined in Part 97?
A. To support teaching of amateur radio classes in schools
B. To provide a voluntary noncommercial communications service to the public, particularly in times of emergency
C. To provide free message service to the public
D. To allow the public to communicate with other radio services

T1A02
(B)
[97.1]
p 5-2

T1A03
What classes of US amateur radio licenses may currently be earned by examination?
A. Novice, Technician, General, Advanced
B. Technician, General, Advanced
C. Technician, General, Extra
D. Technician, Tech Plus, General

T1A03
(C)
[97.501]
p 5-4

T1A04
Who is a Volunteer Examiner?
A. A certified instructor who volunteers to examine amateur teaching manuals
B. An FCC employee who accredits volunteers to administer amateur license exams
C. An amateur accredited by one or more VECs who volunteers to administer amateur license exams
D. Any person who volunteers to examine amateur station equipment

T1A04
(C)
[97.509(b)]
p 5-4

T1A05
(A)
[97.505(a)(6)]
p 5-7

T1A05
How long is a CSCE valid for license upgrade purposes?
A. 365 days
B. Until the current license expires
C. Indefinitely
D. Until two years following the expiration of the current license

T1A06
(D)
[97.509
(a)(b)(3)(i)]
p 5-4

T1A06
How many and what class of Volunteer Examiners are required to administer an Element 2 Technician written exam?
A. Three Examiners holding any class of license
B. Two Examiners holding any class of license
C. Three Examiners holding a Technician Class license
D. Three Examiners holding a General Class license or higher

T1A07
(B)
[97.5]
p 5-2

T1A07
Who makes and enforces the rules for the Amateur Radio Service in the United States?
A. The Congress of the United States
B. The Federal Communications Commission
C. The Volunteer Examiner Coordinators
D. The Federal Bureau of Investigation

T1A08
(D)
[97.1]
p 5-2

T1A08
What are two of the five fundamental purposes for the Amateur Radio Service?
A. To protect historical radio data, and help the public understand radio history
B. To aid foreign countries in improving radio communications and encourage visits from foreign hams
C. To modernize radio electronic design theory and improve schematic drawings
D. To increase the number of trained radio operators and electronics experts, and improve international goodwill

T1A09
(D)
[97.3(a)(5)]
p 5-3

T1A09
What is the definition of an amateur radio station?
A. A station in a public radio service used for radio communications
B. A station using radio communications for a commercial purpose
C. A station using equipment for training new broadcast operators and technicians
D. A station in an Amateur Radio Service consisting of the apparatus necessary for carrying on radio communications

T1A10
(B)
[97.3(A)(23)]
p 6-6

T1A10
What is a transmission called that disturbs other communications?
A. Interrupted CW
B. Harmful interference
C. Transponder signals
D. Unidentified transmissions

T1B - ITU regions, international regulations, US call sign structure, special event calls, vanity call signs - 1 exam question

T1B01
What is the ITU?
A. The International Telecommunications Utility
B. The International Telephone Union
C. The International Telecommunication Union
D. The International Technology Union

T1B01
(C)
[97.3(a)(28)]
p 5-16

T1B02
What is the purpose of ITU Regions?
A. They are used to assist in the management of frequency allocations
B. They are useful when operating maritime mobile
C. They are used in call sign assignments
D. They must be used after your call sign to indicate your location

T1B02
(A)
[97.301
p 5-16

T1B03
What system does the FCC use to select new amateur radio call signs?
A. Call signs are assigned in random order
B. The applicant is allowed to pick a call sign
C. Call signs are assigned in sequential order
D. Volunteer Examiners choose an unassigned call sign

T1B03
(C)
[97.17(d)]
p 5-19

T1B04
What FCC call sign program might you use to obtain a call sign containing your initials?
A. The vanity call sign program
B. The sequential call sign program
C. The special event call sign program
D. There is no FCC provision for choosing a your call sign

T1B04
(A)
[97.19(d)]
p 5-20

T1B05
How might an amateur radio club obtain a club station call sign?
A. By applying directly to the FCC in Gettysburg, PA
B. By applying through a Club Station Call Sign Administrator
C. By submitting a FCC Form 605 to the FCC in Washington, DC
D. By notifying a VE team using NCVEC Form 605

T1B05
(B)
[97.17(b)(2)]
p 5-20

T1B06
Who is eligible to apply for temporary use of a 1-by-1 format Special Event call sign?
A. Only Amateur Extra class amateurs
B. Only military stations
C. Any FCC-licensed amateur
D. Only trustees of amateur radio club stations

T1B06
(C)
p 5-20

T1B07
When are you allowed to operate your amateur station in a foreign country?
A. When there is a reciprocal operating agreement between the countries
B. When there is a mutual agreement allowing third party communications
C. When authorization permits amateur communications in a foreign language
D. When you are communicating with non-licensed individuals in another country

T1B07
(A)
[97.107]
p 5-17

T1B08
(C)
p 5-19

T1B08
Which of the following call signs is a valid US amateur call?
A. UZ4FWD
B. KBL7766
C. KB3TMJ
D. VE3TWJ

T1B09
(B)
p 5-18

T1B09
What letters must be used for the first letter in US amateur call signs?
A. K, N, U and W
B. A, K, N and W
C. A, B, C and D
D. A, N, V and W

T1B10
(D)
p 5-18

T1B10
What numbers are used in US amateur call signs?
A. Any two-digit number, 10 through 99
B. Any two-digit number, 22 through 45
C. A single digit, 1 though 9
D. A single digit, 0 through 9

T1C – Authorized frequencies (Technician), reciprocal licensing, operation near band edges, spectrum sharing – 1 exam question

T1C01
(C)
[97.5(a)]
p 6-1

T1C01
What is required before you can control an amateur station in the US?
A. You must hold an FCC restricted operator's permit for a licensed radio station
B. You must submit an FCC Form 605 with a license examination fee
C. You must be named in the FCC amateur license database, or be an alien with reciprocal operating authorization
D. The FCC must issue you a Certificate of Successful Completion of Amateur Training

T1C02
(B)
[97.5(a)]
p 5-2

T1C02
Where does a US amateur license allow you to transmit?
A. From anywhere in the world
B. From wherever the Amateur Radio Service is regulated by the FCC or where reciprocal agreements are in place
C. From a country that shares a third party agreement with the US
D. Only from the mailing address printed on your license

T1C03
(B)
[97.111]
p 6-13

T1C03
Under what conditions are amateur stations allowed to communicate with stations operating in other radio services?
A. When other radio services make contact with amateur stations
B. When authorized by the FCC
C. When communicating with stations in the Family Radio Service
D. When commercial broadcast stations are off the air

T1C04
(B)
[97.301(a)]
p 5-10

T1C04
Which frequency is within the 6-meter band?
A. 49.00 MHz
B. 52.525 MHz
C. 28.50 MHz
D. 222.15 MHz

T1C05
Which amateur band are you using when transmitting on 146.52 MHz?
A. 2 meter band
B. 20 meter band
C. 14 meter band
D. 6 meter band

T1C05
(A)
[97.301(a)]
p 5-10

T1C06
Which 70-centimeter frequency is authorized to a Technician class license holder operating in ITU Region 2?
A. 455.350 MHz
B. 146.520 MHz
C. 443.350 MHz
D. 222.520 MHz

T1C06
(C)
[97.301(a)]
p 5-10

T1C07
Which 23 centimeter frequency is authorized to a Technician class license holder operating in ITU Region 2?
A. 2315 MHz
B. 1296 MHz
C. 3390 MHz
D. 146.52 MHz

T1C07
(B)
[97.301(a)]
p 5-10

T1C08
What amateur band are you using if you are operating on 223.50 MHz?
A. 15 meter band
B. 10 meter band
C. 2 meter band
D. 1.25 meter band

T1C08
(D)
[97.301(a)]
p 5-10

T1C09
What do the FCC rules mean when an amateur frequency band is said to be available on a secondary basis?
A. Secondary users of a frequency have equal rights to operate
B. Amateurs are only allowed to use the frequency at night
C. Amateurs may not cause harmful interference to primary users
D. Secondary users are not allowed on amateur bands

T1C09
(C)
[97.303]
p 5-15

T1C10
When may a US amateur operator communicate with an amateur in a foreign country?
A. Only when a third-party agreement exists between the US and the foreign country
B. At any time except between 146.52 and 146.58 MHz
C. Only when a foreign amateur uses English
D. At any time unless prohibited by either government

T1C10
(D)
[97.111]
p 5-17

T1C11
Which of the following types of communications are not permitted in the Amateur Radio Service?
A. Brief transmissions to make adjustments to the station
B. Brief transmissions to establish two-way communications with other stations
C. Transmissions to assist persons learning or improving proficiency in CW
D. Communications on a regular basis that could reasonably be furnished alternatively through other radio services

T1C11
(D)
[97.113(a)(5)]
p 6-11

T1D - The station license, correct name and address on file, license term, renewals, grace period – 1 exam question

T1D01
(B)
[97.17(a)]
p 1-14

T1D01

Which of the following services are issued an operator station license by the FCC?
A. Family Radio Service
B. Amateur Radio Service
C. General Radiotelephone Service
D. The Citizens Radio Service

T1D02
(A)
[97.5(b)(1)]
p 5-3

T1D02

Who can become an amateur licensee in the US?
A. Anyone except a representative of a foreign government
B. Only a citizen of the United States
C. Anyone except an employee of the US government
D. Anyone

T1D03
(D)
[97.5(b)(1)]
p 5-3

T1D03

What is the minimum age required to hold an amateur license?
A. 14 years or older
B. 18 years or older
C. 70 years or younger
D. There is no minimum age requirement

T1D04
(D)
[97.5(a)]
p 5-2

T1D04

What government agency grants your amateur radio license?
A. The Department of Defense
B. The Bureau of Public Communications
C. The Department of Commerce
D. The Federal Communications Commission

T1D05
(C)
[97.5(a)]
p 5-7

T1D05

How soon may you transmit after passing the required examination elements for your first amateur radio license?
A. Immediately
B. 30 days after the test date
C. As soon as your license grant appears in the FCC's ULS database
D. As soon as you receive your license in the mail from the FCC

T1D06
(C)
[97.25(a)]
p 5-7

T1D06

What is the normal term for an amateur station license grant?
A. 5 years
B. 7 years
C. 10 years
D. For the lifetime of the licensee

T1D07
(A)
[97.21(b)]
p 5-7

T1D07

What is the grace period during which the FCC will renew an expired 10-year license without re-examination?
A. 2 years
B. 5 years
C. 10 years
D. There is no grace period

T1D08
What is your responsibility as a station licensee?
A. You must allow another amateur to operate your station upon request
B. You must be present whenever the station is operated
C. You must notify the FCC if another amateur acts as the control operator
D. Your station must be operated in accordance with the FCC rules

T1D08
(D)
[97.103(a)]
p 5-8

T1D09
When may the FCC revoke or suspend a license if the mailing address of the holder is not current with the FCC?
A. If mail is returned to the FCC as undeliverable
B. When the licensee transmits without having updated the address
C. When the licensee operates portable at a different address
D. If the address is not updated within the 2 year grace period

T1D09
(A)
[97.23]
p 5-8

T1D10
The FCC requires which address to be kept up to date on the Universal Licensing System database?
A. The station location address
B. The station licensee mailing address
C. The station location address and mailing address
D. The station transmitting location address

T1D10
(B)
[97.23]
p 5-8

T1D11
When are you permitted to continue to transmit if you forget to renew your amateur license and it expires?
A. Transmitting is not allowed until the license is renewed and appears on the FCC ULS database
B. When you identify using the suffix EXP
C. When you notify the FCC you intend to renew within 90 days
D. Transmitting is allowed any time during the 2-year grace period

T1D11
(A)
[97.21(b)]
p 5-7

T1D12
Why must an Amateur radio operator have a correct name and mailing address on file with the FCC?
A. To receive mail delivery from the FCC by the United States Postal Service
B. So the FCC Field office can contact the licensee
C. It isn't required when you haven't operated your station in a year
D. So the FCC can locate your transmitting location

T1D12
(A)
[97.23]
p 5-8

SUBELEMENT T2
Control operator duties
4 exam questions – 4 groups

T2A - Prohibited communications: music, broadcasting, codes and ciphers, business use, permissible communications, bulletins, code practice, incidental music – 1 exam question

T2A01
(A)
[97.113(b)]
p 6-12

T2A01

When is an amateur station authorized to transmit information to the general public?

A. Never
B. Only when the operator is being paid
C. Only when the transmission lasts more than 10 minutes
D. Only when the transmission lasts longer than 15 minutes

T2A02
(A)
[97.113(a)(4),
97.113(e)]
p 6-12

T2A02

When is an amateur station authorized to transmit music?

A. Amateurs may not transmit music, except as incidental to an authorized rebroadcast of space shuttle communications
B. Only when the music produces no spurious emissions
C. Only to interfere with an illegal transmission
D. Only when the music is above 1280 MHz

T2A03
(C)
[97.113(a)(4),
97.211(b),
97.217]
p 6-12

T2A03

When is the transmission of codes or ciphers allowed to hide the meaning of a message transmitted by an amateur station?

A. Only during contests
B. Only when operating mobile
C. Only when transmitting control commands to space stations or radio control craft
D. Only when frequencies above 1280 MHz are used

T2A04
(A)
[97.113(a)(4)]
p 6-11

T2A04

When may an amateur station transmit false or deceptive signals?

A. Never
B. When operating a beacon transmitter in a "fox hunt" exercise
C. Only when making unidentified transmissions
D. When needed to hide the meaning of a message for secrecy

T2A05
(C)
[97.119(b)]
p 6-4, 6-11

T2A05

When may an amateur station transmit unidentified communications?

A. Only during brief tests not meant as messages
B. Only when they do not interfere with others
C. Only when sent from a space station or to control a model craft
D. Only during two-way or third party communications

T2A06
(A)
[97.3(a)(10)]
p 6-12

T2A06

What does the term broadcasting mean?

A. Transmissions intended for reception by the general public, either direct or relayed
B. Retransmission by automatic means of programs or signals from non-amateur stations
C. One-way radio communications, regardless of purpose or content
D. One-way or two-way radio communications between two or more stations

T2A07
Which of the following are specifically prohibited in the Amateur Radio Service?
A. Discussion of politics
B. Discussion of programs on broadcast stations
C. Indecent and obscene language
D. Morse code practice

T2A07
(C)
[97.113(a)(4)]
p 6-11

T2A08
Which of the following one-way communications may not be transmitted in the Amateur Radio Service?
A. Telecommand of model craft
B. Broadcasts intended for reception by the general public
C. Brief transmissions to make adjustments to the station
D. Morse code practice

T2A08
(B)
[97.3(a)(10),
97.113(b)]
p 6-12

T2A09
When does the FCC allow an amateur radio station to be used as a method of communication for hire or material compensation?
A. Only when making test transmissions
B. Only when news is being broadcast in times of emergency
C. Only when in accordance with part 97 rules
D. Only when your employer is using amateur radio to broadcast advertising

T2A09
(C)
[97.113(2)]
p 6-11

T2A10
What type of communications are prohibited when using a repeater autopatch?
A. Calls to a recorded weather report
B. Calls to your employer requesting directions to a customer's office
C. Calls to the police reporting a traffic accident
D. Calls to a public utility reporting an outage of your telephone

T2A10
(B)
[97.113(a)(3),
(a)5(e)]
p 6-11

T2A11
When may you use your station to tell people about equipment you have for sale?
A. Never
B. When you are conducting an on-line auction
C. When you are offering amateur radio equipment for sale or trade on an occasional basis
D. When you are helping a recognized charity

T2A11
(C)
[97.113(a)3]
p 6-11

T2B - Basic identification requirements, repeater ID standards, identification for non-voice modes, identification requirements for mobile and portable operation – 1 exam question

T2B01
What must you transmit to identify your amateur station?
A. Your tactical ID
B. Your call sign
C. Your first name and your location
D. Your full name

T2B01
(B)
[97.119(a)]
p 6-3

T2B02
What is a transmission called that does not contain a station identification?
A. Unidentified communications or signals
B. Reluctance modulation
C. Test emission
D. Intentional interference

T2B02
(A)
[97.119(a)]
p 6-3

T2B03
(B)
[97.119(a)]
p 6-3

T2B03
How often must an amateur station transmit the assigned call sign?
A. At the beginning of each transmission and every 10 minutes during communication
B. Every 10 minutes during communications and at the end of each communication
C. At the end of each transmission
D. Only at the end of the communication

T2B04
(D)
[97.119(b)]
p 6-4

T2B04
What is an acceptable method of transmitting a repeater station identification?
A. By phone using the English language
B. By video image conforming to applicable standards
C. By Morse code at a speed not to exceed 20 words per minute
D. All of these answers are correct.

T2B05
(C)
[97.119(a)]
p 6-3

T2B05
What identification is required when two amateur stations end communications?
A. No identification is required
B. One of the stations must transmit both stations' call signs
C. Each station must transmit its own call sign
D. Both stations must transmit both call signs

T2B06
(B)
[97.119(a)]
p 6-3

T2B06
What is the longest period of time an amateur station can operate without transmitting its call sign?
A. 5 minutes
B. 10 minutes
C. 15 minutes
D. 30 minutes

T2B07
(C)
[97.119(b)(2)]
p 6-3

T2B07
What is a permissible way to identify your station when you are speaking to another amateur operator using a language other than English?
A. You must identify using the official version of the foreign language
B. Identification is not required when using other languages
C. You must identify using the English language
D. You must identify using phonetics

T2B08
(D)
[97.119(d)]
p 5-20, 6-5

T2B08
How often must you identify using your assigned call sign when operating while using a special event call sign?
A. Every 10 minutes
B. Once when the event begins and once when it concludes
C. Never
D. Once per hour

T2B09
(A)
[97.119(4)(c)]
p 5-19

T2B09
What is required when using one or more self-assigned indicators with your assigned call sign?
A. The indicator must not conflict with an indicator specified by FCC rules or with a prefix assigned to another country
B. The indicator must consist only of numeric digits
C. The indicator must include the 2-letter abbreviation for your state
D. The indicator must be separated from your call sign by a double slash mark

T2B10

What is the correct way to identify when visiting a station if you hold a higher class license than that of the station licensee and you are using a frequency not authorized to his class of license?

A. Send your call sign first, followed by his call sign
B. Send his call sign first, followed by your call sign
C. Send your call sign only, his is not required
D. Send his call sign followed by "/KT"

T2B10
(B)
[97.119(e)]
p 6-4

T2B11

When exercising the operating privileges earned by examination upgrade of a license what is meant by use of the indicator "/AG"?

A. Authorized General
B. Adjunct General
C. Address as General
D. Automatically General

T2B11
(A)
[97.119(f)(2)]
p 5-20

T2C – Definition of control operator, location of control operator, automatic and remote control, auxiliary stations – 1 exam question

T2C01

What must every amateur station have when transmitting?

A. A frequency-measuring device
B. A control operator
C. A beacon transmitter
D. A third party operator

T2C01
(B)
[97.7]
p 6-1

T2C02

How many amateur operator / primary station licenses may be held by one person?

A. As many as desired
B. One for each portable transmitter
C. Only one
D. One for each station location

T2C02
(C)
[97.5(b)(1)]
p 5-3

T2C03

What minimum class of amateur license must you hold to be a control operator of a repeater station?

A. Technician Plus
B. Technician
C. General
D. Amateur Extra

T2C03
(B)
[97.205(a)]
p 6-1

T2C04

Who is responsible for the transmissions from an amateur station?

A. Auxiliary operator
B. Operations coordinator
C. Third-party operator
D. Control operator

T2C04
(D)
[97.3(a)
(1)(2)]
p 6-1

T2C05

When must an amateur station have a control operator?

A. Only when training another amateur
B. Whenever the station receiver is operated
C. Whenever the station is transmitting
D. A control operator is not needed

T2C05
(C)
[97.7]
p 6-1, 6-10

T2C06
(D)
[97.3]
p 6-1

T2C06
What is the control point of an amateur station?
A. The on/off switch of the transmitter
B. The input/output port of a packet controller
C. The variable frequency oscillator of a transmitter
D. The location at which the control operator function is performed

T2C07
(C)
[97.109(d)]
p 6-10

T2C07
What type of amateur station does not require a control operator to be at the control point?
A. A locally controlled station
B. A remotely controlled station
C. An automatically controlled station
D. An earth station controlling a space station

T2C08
(A)
[97.3(a)]
p 6-10

T2C08
What are the three types of station control permitted and recognized by FCC rule?
A. Local, remote and automatic control
B. Local, distant and automatic control
C. Remote, distant and unauthorized control
D. All of the choices are correct

T2C09
(C)
[97.3(a)]
p 6-10

T2C09
What type of control is being used on a repeater when the control operator is not present?
A. Local control
B. Remote control
C. Automatic control
D. Uncontrolled

T2C10
(D)
[97.109(a)]
p 6-10

T2C10
What type of control is being used when transmitting using a handheld radio?
A. Radio control
B. Unattended control
C. Automatic control
D. Local control

T2C11
(B)
[97.3]
p 6-10

T2C11
What type of control is used when the control operator is not at the station location but can still make changes to a transmitter?
A. Local control
B. Remote control
C. Automatic control
D. Uncontrolled

T2C12
(C)
[97.3(a)(13)]
p 6-1

T2C12
What is the definition of a control operator of an amateur station?
A. Anyone who operates the controls of the station
B. Anyone who is responsible for the station's equipment
C. An operator designated by the licensee to be responsible for the station's transmissions to assure compliance with FCC rules
D. The operator with the highest class of license who is in control of the station

T2D – Operating another person's station, guest operators at your station, third party communications, autopatch, incidental business use, compensation of operators, club stations, station security, station inspection, protection against unauthorized transmissions – 1 exam question

T2D01
Who is responsible for proper operation if you transmit from another amateur's station?
A. Both of you
B. Only the other station licensee
C. Only you as the control operator
D. Only the station licensee, unless the station records shows another control operator at the time

T2D01
(A)
[97.103(a)]
p 6-2

T2D02
What operating privileges are allowed when another amateur holding a higher class license is controlling your station?
A. All privileges allowed by the higher class license
B. Only the privileges allowed by your license
C. All the emission privileges of the higher class license, but only the frequency privileges of your license
D. All the frequency privileges of the higher class license, but only the emission privileges of your license

T2D02
(A)
[97.105(b)
p 6-2

T2D03
What operating privileges are allowed when you are the control operator at the station of another amateur who has a higher class license than yours?
A. Any privileges allowed by the higher class license
B. Only the privileges allowed by your license
C. All the emission privileges of the higher class license, but only the frequency privileges of your license
D. All the frequency privileges of the higher class license, but only the emission privileges of your license

T2D03
(B)
[97.105(a)]
p 6-2

T2D04
Which of the following is a prohibited amateur radio transmission?
A. Using amateur radio to seek emergency assistance
B. Using amateur radio for conducting business
C. Using an amateur phone patch to call for a taxi or food delivery
D. Using an amateur phone patch to call home to say you are running late

T2D04
(B)
[97.113
(a)(3)]
p 6-11

T2D05
What is the definition of third-party communications?
A. A message sent between two amateur stations for someone else
B. Public service communications for a political party
C. Any messages sent by amateur stations
D. A three-minute transmission to another amateur

T2D05
(A)
[97.3(a)46]
p 6-8

T2D06
How many persons are required to be members of a club for a club station license to be issued by the FCC?
A. At least 5
B. At least 4
C. A trustee and 2 officers
D. At least 2

T2D06
(B)
[97.5(b)(2)]
p 5-20

T2D07
(C)
[97.11(a)]
p 6-12

T2D07
When may you operate your amateur station aboard an aircraft?
A. At any time
B. Only while the aircraft is on the ground
C. Only with the approval of the pilot in command and not using the aircraft's radio equipment
D. Only when you have written permission from the airline and only using the aircraft's radio equipment

T2D08
(B)
[97.103(c)]
p 5-8

T2D08
When is the FCC allowed to inspect your station equipment and station records?
A. Only on weekends
B. At any time upon request
C. Never
D. Only during daylight hours

T2D09
(A)
p 5-8

T2D09
How might you best keep unauthorized persons from using your amateur station?
A. Disconnect the power and microphone cables when not using your equipment
B. Connect a dummy load to the antenna
C. Put a "Danger - High Voltage" sign in the station
D. Put fuses in the main power line

T2D10
(B)
[97.109(b)]
p 5-8

T2D10
Why are unlicensed persons in your family not allowed to transmit on your amateur station if you are not there?
A. They must not use your equipment without your permission
B. They must be licensed before they are allowed to be control operators
C. They must know how to use proper procedures and Q signals
D. They must know the right frequencies and emissions for transmitting

T2D11
(D)
[97.113(d)]
p 6-12

T2D11
When is it permissible for the control operator of a club station to accept compensation for sending information bulletins or Morse code practice?
A. When compensation is paid from a non-profit organization
B. When the club station license is held by a non-profit organization
C. Anytime compensation is needed
D. When the station makes those transmissions for at least 40 hours per week

SUBELEMENT T3
Operating practices – 4 exam questions – 4 groups

T3A - Choosing an operating frequency, calling CQ, calling another station, test transmissions – 1 exam question

T3A01
Which of the following should you do when selecting a frequency on which to transmit?
A. Call CQ to see if anyone is listening
B. Listen to determine if the frequency is busy
C. Transmit on a frequency that allows your signals to be heard
D. Check for maximum power output

T3A01
(B)
p 4-12

T3A02
How do you call another station on a repeater if you know the station's call sign?
A. Say "break, break" then say the station's call sign
B. Say the station's call sign then identify your own station
C. Say "CQ" three times then the other station's call sign
D. Wait for the station to call "CQ" then answer it

T3A02
(B)
p 4-12

T3A03
How do you indicate you are looking for any station with which to make contact?
A. CQ followed by your callsign
B. RST followed by your callsign
C. QST followed by your callsign
D. SK followed by your callsign

T3A03
(A)
p 4-12

T3A04
What should you transmit when responding to a call of CQ?
A. Your own CQ followed by the other station's callsign
B. Your callsign followed by the other station's callsign
C. The other station's callsign followed by your callsign
D. A signal report followed by your callsign

T3A04
(C)
p 4-12

T3A05
What term describes a brief test transmission that does not include any station identification?
A. A test emission with no identification required
B. An illegal un-modulated transmission
C. An illegal unidentified transmission
D. A non-voice ID transmission

T3A05
(C)
[97.119(a)]
p 6-3

T3A06
What must an amateur do when making a transmission to test equipment or antennas?
A. Properly identify the station
B. Make test transmissions only after 10:00 PM local time
C. Notify the FCC of the test transmission
D. State the purpose of the test during the test procedure

T3A06
(A)
p 6-4

T3A07
Which of the following is true when making a test transmission?
A. Station identification is not required if the transmission is less than 15 seconds
B. Station identification is not required if the transmission is less than 1 watt
C. Station identification is required only if your station can be heard
D. Station identification is required at least every ten minutes and at the end of every transmission.

T3A07
(D)
p 6-4

T3A08
(D)
p 4-12

T3A08
What is the meaning of the procedural signal "CQ"?
A. Call on the quarter hour
B. New antenna is being tested (no station should answer)
C. Only the called station should transmit
D. Calling any station

T3A09
(A)
[97.119(b)(2)]
p 4-13

T3A09
Why should you avoid using cute phrases or word combinations to identify your station?
A. They are not easily understood by some operators
B. They might offend some operators
C. They do not meet FCC identification requirements
D. They might be interpreted as codes or ciphers intended to obscure your identification

T3A10
(B)
p 4-12

T3A10
What brief statement is often used in place of "CQ" to indicate that you are listening for calls on a repeater?
A. Say "Hello test" followed by your call sign
B. Say your call sign
C. Say the repeater call sign followed by your call sign
D. Say the letters "QSY" followed by your call sign

T3A11
(A)
[97.119(b)(2)]
p 4-13

T3A11
Why should you use the International Telecommunication Union (ITU) phonetic alphabet when identifying your station?
A. The words are internationally recognized substitutes for letters
B. There is no advantage
C. The words have been chosen to represent amateur radio terms
D. It preserves traditions begun in the early days of amateur radio

T3B Use of minimum power, band plans, repeater coordination, mode restricted sub-bands 1 exam question

T3B01
What is a band plan?
A. A voluntary guideline, beyond the divisions established by the FCC for using different operating modes within an amateur band
B. A guideline from the FCC for making amateur frequency band allocations
C. A guideline for operating schedules within an amateur band published by the FCC
D. A plan devised by a local group

T3B01
(A)
p 4-9

T3B02
Which of the following statements is true of band plans?
A. They are mandated by the FCC to regulate spectrum use
B. They are mandated by the ITU
C. They are voluntary guidelines for efficient use of the radio spectrum
D. They are mandatory only in the US

T3B02
(C)
p 4-10

T3B03
Who developed the band plans used by amateur radio operators?
A. The US Congress
B. The FCC
C. The amateur community
D. The Interstate Commerce Commission

T3B03
(C)
p 4-11

T3B04
Who is in charge of the repeater frequency band plan in your local area?
A. The local FCC field office
B. RACES and FEMA
C. The recognized frequency coordination body
D. Repeater Council of America

T3B04
(C)
p 4-17

T3B05
What is the main purpose of repeater coordination?
A. To reduce interference and promote proper use of spectrum
B. To coordinate as many repeaters as possible in a small area
C. To coordinate all possible frequencies available for repeater use
D. To promote and encourage use of simplex frequencies

T3B05
(A)
p 4-17

T3B06
Who is accountable if a repeater station inadvertently retransmits communications that violate FCC rules?
A. The repeater trustee
B. The repeater control operator
C. The transmitting station
D. All of these answers are correct

T3B06
(C)
[97.205(g)]
p 4-16

T3B07
Which of these statements is true about legal power levels on the amateur bands?
A. Always use the maximum power allowed to ensure that you complete the contact
B. An amateur may use no more than 200 Watts PEP to make an amateur contact
C. An amateur may use up to 1500 Watts PEP on any amateur frequency
D. An amateur must use the minimum transmitter power necessary to carry out the desired communication

T3B07
(D)
p 4-3

T3B08
(C)
[97.305(c)]
p 5-13

T3B08
Which of the bands available to Technician class licensees have mode restricted sub-bands?
A. The 6-meter, 2-meter, and 70-centimeter bands
B. The 2-meter and 13-centimeter bands
C. The 6-meter, 2-meter, and 1¼-meter bands
D. The 2-meter and 70-centimeter bands

T3B09
(A)
[97.305
(a)(c)]
p 5-13

T3B09
What emission modes are permitted in the restricted sub-band at
50.0-50.1 MHz?
A. CW only
B. CW and RTTY
C. SSB only
D. CW and SSB

T3B10
(A)
[97.305
(a)(c)]
p 5-13

T3B10
What emission modes are permitted in the restricted sub-band at
144.0-144.1 MHz?
A. CW only
B. CW and RTTY
C. SSB only
D. CW and SSB

T3B11

T3B11
This question has been withdrawn.

T3C - Courtesy and respect for others, sensitive subject areas, obscene and indecent language – 1 exam question

T3C01
What is the proper way to break into a conversation between two stations that are using the frequency?
A. Say your call sign between their transmissions
B. Wait for them to finish and then call CQ
C. Say "Break-break" between their transmissions
D. Call one of the operators on the telephone to interrupt the conversation

T3C02
What is considered to be proper repeater operating practice?
A. Monitor before transmitting and keep transmissions short
B. Identify legally
C. Use the minimum amount of transmitter power necessary
D. All of these answers are correct

T3C03
What should you do before responding to another stations call?
A. Make sure you are operating on a permissible frequency for your license class
B. Adjust your transmitter for maximum power output
C. Ask the station to send their signal report and location
D. Verify the other station's license class

T3C04
What rule applies if two amateur stations want to use the same frequency?
A. The station operator with a lesser class of license must yield the frequency to a higher-class licensee
B. The station operator with a lower power output must yield the frequency to the station with a higher power output
C. No frequency will be assigned for the exclusive use of any station and neither has priority
D. Station operators in ITU Regions 1 and 3 must yield the frequency to stations in ITU Region 2

T3C05
Why is indecent and obscene language prohibited in the Amateur Service?
A. Because it is offensive to some individuals
B. Because young children may intercept amateur communications with readily available receiving equipment
C. Because such language is specifically prohibited by FCC Rules
D. All of these choices are correct

T3C06
Why should amateur radio operators avoid the use of racial or ethnic slurs when talking to other stations?
A. Such language is prohibited by the FCC
B. It is offensive to some people and reflects a poor public image on all amateur radio operators
C. Some of the terms used may be unfamiliar to other operators
D. You transmissions might be recorded for use in court

T3C01
(A)
p 4-14

T3C02
(D)
p 4-16

T3C03
(A)
p 4-12

T3C04
(C)
[97.101(b)]
p 4-2

T3C05
(D)
[97.113(a)(4)]
p 4-5

T3C06
(B)
p 4-5

T3C07
(C)
p 4-6

T3C07
What should you do if you hear a newly licensed operator that is having trouble with their station?
A. Tell them to get off the air until they learn how operate properly
B. Report them to the FCC
C. Contact them and offer to help with the problem
D. Move to another frequency

T3C08
(B)
[97.113(a)(4)]
p 4-5

T3C08
Where can an official list be found of prohibited obscene and indecent words that should not be used in amateur radio?
A. On the FCC web site
B. There is no official list of prohibited obscene and indecent words
C. On the Department of Commerce web site
D. The official list is in public domain and found in all amateur study guides

T3C09
(D)
[97.113(a)(4)]
p 4-5

T3C09
What type of subjects are not prohibited communications while using amateur radio?
A. Political discussions
B. Jokes and stories
C. Religious preferences
D. All of these answers are correct

T3C10
(C)
[97.101 (a)]
p 4-1

T3C10
When circumstances are not specifically covered by FCC rules what general operating standard must be applied to amateur station operation?
A. Designated operator control
B. Politically correct control
C. Good engineering and amateur practices
D. Reasonable operator control

T3D - Interference to and from consumer devices, public relations, intentional and unintentional interference - 1 exam question

T3D01
What should you do if you receive a report that your transmissions are causing splatter or interference on nearby frequencies?
A. Increase transmit power
B. Change mode of transmission
C. Report the interference to the equipment manufacturer
D. Check transmitter for off frequency operation or spurious emissions

T3D02
Who is responsible for taking care of the interference if signals from your transmitter are causing front end overload in your neighbor's television receiver?
A. You alone are responsible, since your transmitter is causing the problem
B. Both you and the owner of the television receiver share the responsibility
C. The FCC must decide if you or the owner of the television receiver is responsible
D. The owner of the television receiver is responsible

T3D03
What is the major cause of telephone interference?
A. The telephone wiring is inadequate
B. Tropospheric ducting at UHF frequencies
C. The telephone was not equipped with adequate interference protection when manufactured.
D. Improper location of the telephone in the home

T3D04
What is the proper course of action if you unintentionally interfere with another station?
A. Rotate your antenna slightly
B. Properly identify your station and move to a different frequency
C. Increase power
D. Change antenna polarization

T3D05
When may you deliberately interfere with another station's communications?
A. Only if the station is operating illegally
B. Only if the station begins transmitting on a frequency you are using
C. Never
D. You may cause deliberate interference because it can't be helped during crowded band conditions

T3D06
Who has exclusive use of a specific frequency when the FCC has not declared a communication emergency?
A. Any net station that has traffic
B. The station first occupying the frequency
C. Individuals passing health and welfare communications
D. No station has exclusive use of any frequency

T3D01
(D)
p 6-6

T3D02
(D)
p 3-31

T3D03
(C)
p 3-31

T3D04
(B)
p 6-7

T3D05
(C)
[97.101(d)]
p 6-7

T3D06
(D)
p 4-14, 4-25

T3D07
(C)
p 3-32

T3D07
What effect might a break in a cable television transmission line have on amateur communications?
A. A break cannot affect amateur communications
B. Harmonic radiation from the TV may cause the amateur transmitter to transmit off-frequency
C. TV interference may result when the amateur station is transmitting, or interference may occur to the amateur receiver
D. The broken cable may pick up very high voltages when the amateur station is transmitting

T3D08
(C)
p 6-7

T3D08
What is the best way to reduce on the air interference when testing your transmitter?
A. Use a short indoor antenna when testing
B. Use upper side band when testing
C. Use a dummy load when testing
D. Use a simplex frequency instead of a repeater frequency

T3D09
(C)
[97.103(a)]
p 4-24

T3D09
What rules apply to your station when using amateur radio at the request of public service officials or at the scene of an emergency?
A. RACES
B. ARES
C. FCC
D. FEMA

T3D10
(D)
p 4-27

T3D10
What do RACES and ARES have in common?
A. They represent the two largest ham clubs in the United States
B. One handles road traffic, the other weather traffic
C. Neither may handle emergency traffic
D. Both organizations provide communications during emergencies

T3D11
(C)
p 3-31

T3D11
What is meant by receiver front-end overload?
A. Too much voltage from the power supply
B. Too much current from the power supply
C. Interference caused by strong signals from a nearby source
D. Interference caused by turning the volume up too high

SUBELEMENT T4

Radio and electronic fundamentals

5 exam questions – 5 groups

T4A – Names of electrical units, DC and AC, what is a radio signal, conductors and insulators, electrical components - 1 exam question

T4A01
Electrical current is measured in which of the following units?
A. Volts
B. Watts
C. Ohms
D. Amperes

T4A02
Electrical Power is measured in which of the following units?
A. Volts
B. Watts
C. Ohms
D. Amperes

T4A03
What is the name for the flow of electrons in an electric circuit?
A. Voltage
B. Resistance
C. Capacitance
D. Current

T4A04
What is the name of a current that flows only in one direction?
A. An alternating current
B. A direct current
C. A normal current
D. A smooth current

T4A05
What is the standard unit of frequency?
A. The megacycle
B. The Hertz
C. One thousand cycles per second
D. The electromagnetic force

T4A06
How much voltage does an automobile battery usually supply?
A. About 12 volts
B. About 30 volts
C. About 120 volts
D. About 240 volts

T4A07
What is the basic unit of resistance?
A. The volt
B. The watt
C. The ampere
D. The ohm

T4A01
(D)
p 2-4

T4A02
(B)
p 2-5

T4A03
(D)
p 2-4

T4A04
(B)
p 2-6

T4A05
(B)
p 2-15

T4A06
(A)
p 3-26

T4A07
(D)
p 2-5

T4A08
(A)
p 2-6

T4A08
What is the name of a current that reverses direction on a regular basis?
A. An alternating current
B. A direct current
C. A circular current
D. A vertical current

T4A09
(C)
p 2-5

T4A09
Which of the following is a good electrical conductor?
A. Glass
B. Wood
C. Copper
D. Rubber

T4A10
(B)
p 2-5

T4A10
Which of the following is a good electrical insulator?
A. Copper
B. Glass
C. Aluminum
D. Mercury

T4A11
(B)
p 2-5

T4A11
What is the term used to describe opposition to current flow in ordinary conductors such as wires?
A. Inductance
B. Resistance
C. Counter EMF
D. Magnetism

T4A12
(C)
p 2-4

T4A12
What instrument is used to measure the flow of current in an electrical circuit?
A. Frequency meter
B. SWR meter
C. Ammeter
D. Voltmeter

T4A13
(B)
p 2-4

T4A13
What instrument is used to measure Electromotive Force (EMF) between two points such as the poles of a battery?
A. Magnetometer
B. Voltmeter
C. Ammeter
D. Ohmmeter

T4B – relationship between frequency and wavelength, identification of bands, names of frequency ranges, types of waves – 1 exam question

T4B01
What is the name for the distance a radio wave travels during one complete cycle?
A. Wave speed
B. Waveform
C. Wavelength
D. Wave spread

T4B01
(C)
p 2-15

T4B02
What term describes the number of times that an alternating current flows back and forth per second?
A. Pulse rate
B. Speed
C. Wavelength
D. Frequency

T4B02
(D)
p 2-15

T4B03
What does 60 hertz (Hz) mean?
A. 6000 cycles per second
B. 60 cycles per second
C. 6000 meters per second
D. 60 meters per second

T4B03
(B)
p 2-15

T4B04
Electromagnetic waves that oscillate more than 20,000 times per second as they travel through space are generally referred to as what?
A. Gravity waves
B. Sound waves
C. Radio waves
D. Gamma radiation

T4B04
(C)
p 2-17

T4B05
How fast does a radio wave travel through space?
A. At the speed of light
B. At the speed of sound
C. Its speed is inversely proportional to its wavelength
D. Its speed increases as the frequency increases

T4B05
(A)
p 2-15

T4B06
How does the wavelength of a radio wave relate to its frequency?
A. The wavelength gets longer as the frequency increases
B. The wavelength gets shorter as the frequency increases
C. There is no relationship between wavelength and frequency
D. The wavelength depends on the bandwidth of the signal

T4B06
(B)
p 2-16

T4B07
What is the formula for converting frequency to wavelength in meters?
A. Wavelength in meters equals frequency in Hertz multiplied by 300
B. Wavelength in meters equals frequency in Hertz divided by 300
C. Wavelength in meters equals frequency in megahertz divided by 300
D. Wavelength in meters equals 300 divided by frequency in megahertz

T4B07
(D)
p 2-16

T4B08
(C)
p 2-20

T4B08
What are sound waves in the range between 300 and 3000 Hertz called?
A. Test signals
B. Ultrasonic waves
C. Voice frequencies
D. Radio frequencies

T4B09
(A)
p 2-16

T4B09
What property of a radio wave is often used to identify the different bands amateur radio operators use?
A. The physical length of the wave
B. The magnetic intensity of the wave
C. The time it takes for the wave to travel one mile
D. The voltage standing wave ratio of the wave

T4B10
(A)
p 5-10, 5-14

T4B10
What is the frequency range of the 2 meter band in the United States?
A. 144 to 148 MHz
B. 222 to 225 MHz
C. 420 to 450 MHz
D. 50 to 54 MHz

T4B11
(D)
p 5-10, 5-14

T4B11
What is the frequency range of the 6 meter band in the United States?
A. 144 to 148 MHz
B. 222 to 225 MHz
C. 420 to 450 MHz
D. 50 to 54 MHz

T4B12
(C)
p 5-10, 5-14

T4B12
What is the frequency range of the 70 centimeter band in the United States?
A. 144 to 148 MHz
B. 222 to 225 MHz
C. 420 to 450 MHz
D. 50 to 54 MHz

T4C - How radio works: receivers, transmitters, transceivers, amplifiers, power supplies, types of batteries, service life – 1 exam question

T4C01
What is used to convert radio signals into sounds we can hear?
A. Transmitter
B. Receiver
C. Microphone
D. Antenna

T4C02
What is used to convert sounds from our voice into radio signals?
A. Transmitter
B. Receiver
C. Speaker
D. Antenna

T4C03
What two devices are combined into one unit in a transceiver?
A. Receiver, transmitter
B. Receiver, transformer
C. Receiver, transistor
D. Transmitter, deceiver

T4C04
What device is used to convert the alternating current from a wall outlet into low-voltage direct current?
A. Inverter
B. Compressor
C. Power Supply
D. Demodulator

T4C05
What device is used to increase the output of a 10 watt radio to 100 watts?
A. Amplifier
B. Power supply
C. Antenna
D. Attenuator

T4C06
Which of the battery types listed below offers the longest life when used with a hand-held radio, assuming each battery is the same physical size?
A. Lead-acid
B. Alkaline
C. Nickel-cadmium
D. Lithium-ion

T4C07
What is the nominal voltage per cell of a fully charged nickel-cadmium battery?
A. 1.0 volts
B. 1.2 volts
C. 1.5 volts
D. 2.2 volts

T4C01
(B)
p 2-1

T4C02
(A)
p 2-1

T4C03
(A)
p 2-2

T4C04
(C)
p 2-2, 3-24

T4C05
(A)
p 2-3

T4C06
(D)
p 3-25, 3-26

T4C07
(B)
p 3-25

T4C08
(B)
p 3-25

T4C08
What battery type on this list is not designed to be re-charged?
A. Nickel-cadmium
B. Carbon-zinc
C. Lead-acid
D. Lithium-ion

T4C09
(D)
p 3-26

T4C09
What is required to keep rechargeable batteries in good condition and ready for emergencies?
A. They must be inspected for physical damage and replaced if necessary
B. They should be stored in a cool and dry location
C. They must be given a maintenance recharge at least every 6 months
D. All of these answers are correct

T4C10
(B)
p 3-26

T4C10
What is the best way to get the most amount of energy from a battery?
A. Draw current from the battery as rapidly as possible
B. Draw current from the battery at the slowest rate needed
C. Reverse the leads when the battery reaches the $1/2$ charge level
D. Charge the battery as frequently as possible

T4D – Ohms law relationships – 1 exam question

T4D01
What formula is used to calculate current in a circuit?
A. Current (I) equals voltage (E) multiplied by resistance (R)
B. Current (I) equals voltage (E) divided by resistance (R)
C. Current (I) equals voltage (E) added to resistance (R)
D. Current (I) equals voltage (E) minus resistance (R)

T4D02
What formula is used to calculate voltage in a circuit?
A. Voltage (E) equals current (I) multiplied by resistance (R)
B. Voltage (E) equals current (I) divided by resistance (R)
C. Voltage (E) equals current (I) added to resistance (R)
D. Voltage (E) equals current (I) minus resistance (R)

T4D03
What formula is used to calculate resistance in a circuit?
A. Resistance (R) equals voltage (E) multiplied by current (I)
B. Resistance (R) equals voltage (E) divided by current (I)
C. Resistance (R) equals voltage (E) added to current (I)
D. Resistance (R) equals voltage (E) minus current (I)

T4D04
What is the resistance of a circuit when a current of 3 amperes flows through a resistor connected to 90 volts?
A. 3 ohms
B. 30 ohms
C. 93 ohms
D. 270 ohms

T4D05
What is the resistance in a circuit where the applied voltage is 12 volts and the current flow is 1.5 amperes?
A. 18 ohms
B. 0.125 ohms
C. 8 ohms
D. 13.5 ohms

T4D06
What is the current flow in a circuit with an applied voltage of 120 volts and a resistance of 80 ohms?
A. 9600 amperes
B. 200 amperes
C. 0.667 amperes
D. 1.5 amperes

T4D07
What is the voltage across the resistor if a current of 0.5 amperes flows through a 2 ohm resistor?
A. 1 volt
B. 0.25 volts
C. 2.5 volts
D. 1.5 volts

T4D01
(B)
p 2-5

T4D02
(A)
p 2-5

T4D03
(B)
p 2-5

T4D04
(B)
p 2-5

T4D05
(C)
p 2-5

T4D06
(D)
p 2-5

T4D07
(A)
p 2-5

T4D08
(A)
p 2-5

T4D08
What is the voltage across the resistor if a current of 1 ampere flows through a 10 ohm resistor?
A. 10 volts
B. 1 volt
C. 11 volts
D. 9 volts

T4D09
(A)
p 2-5

T4D09
What is the voltage across the resistor if a current of 2 amperes flows through a 10 ohm resistor?
A. 20 volts
B. 0.2 volts
C. 12 volts
D. 8 volts

T4D10
(C)
p 2-5

T4D10
What is the current flowing through a 100 ohm resistor connected across 200 volts?
A. 20,000 amperes
B. 0.5 amperes
C. 2 amperes
D. 100 amperes

T4D11
(C)
p 2-5

T4D11
What is the current flowing through a 24 ohm resistor connected across 240 volts?
A. 24,000 amperes
B. 0.1 amperes
C. 10 amperes
D. 216 amperes

T4E - Power calculations, units, kilo, mega, milli, micro - 1 exam question

T4E01
What unit is used to describe electrical power?
A. Ohm
B. Farad
C. Volt
D. Watt

T4E01
(D)
p 2-5

T4E02
What is the formula used to calculate electrical power in a DC circuit?
A. Power (P) equals voltage (E) multiplied by current (I)
B. Power (P) equals voltage (E) divided by current (I)
C. Power (P) equals voltage (È) minus current (I)
D. Power (P) equals voltage (E) plus current (I)

T4E02
(A)
p 2-5

T4E03
How much power is represented by a voltage of 13.8 volts DC and a current of 10 amperes?
A. 138 watts
B. 0.7 watts
C. 23.8 watts
D. 3.8 watts

T4E03
(A)
p 2-6

T4E04
How much power is being used in a circuit when the voltage is 120 volts DC and the current is 2.5 amperes?
A. 1440 watts
B. 300 watts
C. 48 watts
D. 30 watts

T4E04
(B)
p 2-6

T4E05
How can you determine how many watts are being drawn by your transceiver when you are transmitting?
A. Measure the DC voltage and divide it by 60 Hz
B. Check the fuse in the power leads to see what size it is
C. Look in the Radio Amateur's Handbook
D. Measure the DC voltage at the transceiver and multiply by the current drawn when you transmit

T4E05
(D)
p 2-6

T4E06
How many amperes are flowing in a circuit when the applied voltage is 120 volts DC and the load is 1200 watts?
A. 20 amperes
B. 10 amperes
C. 120 amperes
D. 5 amperes

T4E06
(B)
p 2-6

T4E07
How many milliamperes is the same as 1.5 amperes?
A. 15 milliamperes
B. 150 milliamperes
C. 1500 milliamperes
D. 15000 milliamperes

T4E07
(C)
p 2-7

T4E08
(A)
p 2-7

T4E08
What is another way to specify the frequency of a radio signal that is oscillating at 1,500,000 Hertz?
A. 1500 kHz
B. 1500 MHz
C. 15 GHz
D. 150 kHz

T4E09
(C)
p 2-7

T4E09
How many volts are equal to one kilovolt?
A. one one-thousandth of a volt
B. one hundred volts
C. one thousand volts
D. one million volts

T4E10
(A)
p 2-7

T4E10
How many volts are equal to one microvolt?
A. one one-millionth of a volt
B. one million volts
C. one thousand kilovolt
D. one one-thousandth of a volt

T4E11
(B)
p 2-7

T4E11
How many watts does a hand-held transceiver put out if the output power is 500 milliwatts?
A. 0.02 watts
B. 0.5 watts
C. 5 watts
D. 50 watts

SUBELEMENT T5

Station setup and operation

4 exam questions – 4 groups

T5A - Station hookup – microphone, speaker, headphones, filters, power source, connecting a computer – 1 exam question

T5A01
What does a microphone connect to in a basic amateur radio station?
A. The receiver
B. The transmitter
C. The SWR Bridge
D. The Balun

T5A02
Which piece of station equipment converts electrical signals to sound waves?
A. Frequency coordinator
B. Frequency discriminator
C. Speaker
D. Microphone

T5A03
What is the term used to describe what happens when a microphone and speaker are too close to each other?
A. Excessive wind noise
B. Audio feedback
C. Inverted signal patterns
D. Poor electrical grounding

T5A04
What could you use in place of a regular speaker to help you copy signals in a noisy area?
A. A video display
B. A low pass filter
C. A set of headphones
D. A boom microphone

T5A05
What is a good reason for using a regulated power supply for communications equipment?
A. To protect equipment from voltage fluctuations
B. A regulated power supply has FCC approval
C. A fuse or circuit breaker regulates the power
D. Regulated supplies are less expensive

T5A06
Where must a filter be installed to reduce spurious emissions?
A. At the transmitter
B. At the receiver
C. At the station power supply
D. At the microphone

T5A01	(B) p 2-3
T5A02	(C) p 2-3
T5A03	(B) p 3-4
T5A04	(C) p 2-3
T5A05	(A) p 3-24
T5A06	(A) p 3-30

T5A07
(D)
p 3-30

T5A07
What type of filter should be connected to a TV receiver as the first step in trying to prevent RF overload from a nearby 2-meter transmitter?
A. Low-pass filter
B. High-pass filter
C. Band pass filter
D. Notch filter

T5A08
(C)
p 3-10

T5A08
What is connected between the transceiver and computer terminal in a packet radio station?
A. Transmatch
B. Mixer
C. Terminal Node Controller
D. Antenna

T5A09
(D)
p 3-10

T5A09
Which of these items is not required for a packet radio station?
A. Antenna
B. Transceiver
C. Power source
D. Microphone

T5A10
(B)
p 3-10

T5A10
What can be used to connect a radio with a computer for data transmission?
A. Balun
B. Sound Card
C. Impedance matcher
D. Autopatch

T5B - Operating controls – 1 exam question

T5B01
What may happen if a transmitter is operated with the microphone gain set too high?
A. The output power will be too high
B. It may cause the signal to become distorted and unreadable
C. The frequency will vary
D. The SWR will increase

T5B02
What kind of information may a VHF/UHF transceiver be capable of storing in memory?
A. Transmit and receive operating frequency
B. CTCSS tone frequency
C. Transmit power level
D. All of these answers are correct

T5B03
What is one way to select a frequency on which to operate?
A. Use the keypad or VFO knob to enter the correct frequency
B. Turn on the CTCSS encoder
C. Adjust the power supply ripple frequency
D. All of these answers are correct

T5B04
What is the purpose of the squelch control on a transceiver?
A. It is used to set the highest level of volume desired
B. It is used to set the transmitter power level
C. It is used to adjust the antenna polarization
D. It is used to quiet noise when no signal is being received

T5B05
What is a way to enable quick access to a favorite frequency on your transceiver?
A. Enable the CTCSS tones
B. Store the frequency in a memory channel
C. Disable the CTCSS tones
D. Use the scan mode to select the desired frequency

T5B06
What might you do to improve the situation if the station you are listening to is hard to copy because of ignition noise interference?
A. Increase your transmitter power
B. Decrease the squelch setting
C. Turn on the noise blanker
D. Use the RIT control

T5B07
What is the purpose of the buttons labeled "up" and "down" on many microphones?
A. To allow easy frequency or memory selection
B. To raise or lower the internal antenna
C. To set the battery charge rate
D. To upload or download messages

T5B01
(B)
p 3-5

T5B02
(D)
p 3-2

T5B03
(A)
p 3-2

T5B04
(D)
p 3-7

T5B05
(B)
p 3-2

T5B06
(C)
p 3-7

T5B07
(A)
p 3-3

T5B08
(C)
p 3-9

T5B08
What is the purpose of the "shift" control found on many VHF/UHF transceivers?
A. Adjust transmitter power level
B. Change bands
C. Adjust the offset between transmit and receive frequency
D. Change modes

T5B09
(B)
p 3-7

T5B09
What does RIT mean?
A. Receiver Input Tone
B. Receiver Incremental Tuning
C. Rectifier Inverter Test
D. Remote Input Transmitter

T5B10
(D)
p 3-2

T5B10
What is the purpose of the "step" menu function found on many transceivers?
A. It adjusts the transmitter power output level
B. It adjusts the modulation level
C. It sets the earphone volume
D. It sets the tuning rate when changing frequencies

T5B11
(C)
p 3-3

T5B11
What is the purpose of the "function" or "F" key found on many transceivers?
A. It turns the power on and off
B. It selects the autopatch access code
C. It selects an alternate action for some control buttons
D. It controls access to the memory scrambler

T5C – Repeaters; repeater and simplex operating techniques, offsets, selective squelch, open and closed repeaters, linked repeaters - 1 exam question

T5C01
What is one purpose of a repeater?
A. To cut your power bill by using someone else's higher power system
B. To extend the usable range of mobile and low-power stations
C. To transmit signals for observing propagation and reception
D. To communicate with stations in services other than amateur

T5C01
(B)
p 2-2

T5C02
What is a courtesy tone?
A. A tone used to identify the repeater
B. A tone used to indicate when a transmission is complete
C. A tone used to indicate that a message is waiting for someone
D. A tone used to activate a receiver in case of severe weather

T5C02
(B)
p 4-13

T5C03
Which of the following is the most important information to know before using a repeater?
A. The repeater input and output frequencies
B. The repeater call sign
C. The repeater power level
D. Whether or not the repeater has an autopatch

T5C03
(A)
p 3-8

T5C04
Why should you pause briefly between transmissions when using a repeater?
A. To let your radio cool off
B. To reach for pencil and paper so you can take notes
C. To listen for anyone wanting to break in
D. To dial up the repeater's autopatch

T5C04
(C)
p 4-15

T5C05
What is the most common input/output frequency offset for repeaters in the 2-meter band?
A. 0.6 MHz
B. 1.0 MHz
C. 1.6 MHz
D. 5.0 MHz

T5C05
(A)
p 3-9

T5C06
What is the most common input/output frequency offset for repeaters in the 70-centimeter band?
A. 600 kHz
B. 1.0 MHz
C. 1.6 MHz
D. 5.0 MHz

T5C06
(D)
p 3-9

T5C07
What is meant by the terms input and output frequency when referring to repeater operations?
A. The repeater receives on one frequency and transmits on another
B. The repeater offers a choice of operating frequencies
C. One frequency is used to control the repeater and another is used to retransmit received signals
D. The repeater must receive an access code on one frequency before it will begin transmitting

T5C07
(A)
p 3-8

T5C08
(A)
p 3-8

T5C08
What is the meaning of the term simplex operation?
A. Transmitting and receiving on the same frequency
B. Transmitting and receiving over a wide area
C. Transmitting on one frequency and receiving on another
D. Transmitting one-way communications

T5C09
(B)
p 4-18

T5C09
What is a reason to use simplex instead of a repeater?
A. When the most reliable communications are needed
B. To avoid tying up the repeater when direct contact is possible
C. When an emergency telephone call is needed
D. When you are traveling and need some local information

T5C10
(A)
p 4-18

T5C10
How might you find out if you could communicate with a station using simplex instead of a repeater?
A. Check the repeater input frequency to see if you can hear the other station
B. Check to see if you can hear the other station on a different frequency band
C. Check to see if you can hear a more distant repeater
D. Check to see if a third station can hear both of you

T5C11
(C)
p 4-15

T5C11
What is the term for a series of repeaters that can be connected to one another to provide users with a wider coverage?
A. Open repeater system
B. Closed repeater system
C. Linked repeater system
D. Locked repeater system

T5C12
(A)
p 4-17

T5C12
What is the main reason repeaters should be approved by the local frequency coordinator before being installed?
A. Coordination minimizes interference between repeaters and makes the most efficient use of available frequencies
B. Coordination is required by the FCC
C. Repeater manufacturers have exclusive territories and you could be fined for using the wrong equipment
D. Only coordinated systems will be approved by the officers of the local radio club

T5C13
(B)
p 4-17

T5C13
Which of the following statements regarding use of repeaters is true?
A. All amateur radio operators have the right to use any repeater at any time
B. Access to any repeater may be limited by the repeater owner
C. Closed repeaters must be opened at the request of any amateur wishing to use it
D. Open repeaters are required to use CTCSS tones for access

T5C14
(D)
p 4-17

T5C14
What term is used to describe a repeater when use is restricted to the members of a club or group?
A. A beacon station
B. An open repeater
C. A auxiliary station
D. A closed repeater

T5D – Recognition and correction of problems, symptoms of overload and overdrive, distortion, over and under modulation, RF feedback, off frequency signals, fading and noise, problems with digital communications links – 1 exam question

T5D01
What is meant by fundamental overload in reference to a receiver?
A. Too much voltage from the power supply
B. Too much current from the power supply
C. Interference caused by very strong signals from a nearby source
D. Interference caused by turning the volume up too high

T5D02
Which of the following is NOT a cause of radio frequency interference?
A. Fundamental overload
B. Doppler shift
C. Spurious emissions
D. Harmonics

T5D03
What is the most likely cause of telephone interference from a nearby transmitter?
A. Harmonics from the transmitter
B. The transmitter's signals are causing the telephone to act like a radio receiver
C. Poor station grounding
D. Improper transmitter adjustment

T5D04
What is a logical first step when attempting to cure a radio frequency interference problem in a nearby telephone?
A. Install a low-pass filter at the transmitter
B. Install a high-pass filter at the transmitter
C. Install an RF filter at the telephone
D. Improve station grounding

T5D05
What should you do first if someone tells you that your transmissions are interfering with their TV reception?
A. Make sure that your station is operating properly and that it does not cause interference to your own television
B. Immediately turn off your transmitter and contact the nearest FCC office for assistance
C. Tell them that your license gives you the right to transmit and nothing can be done to reduce the interference
D. Continue operating normally because your equipment cannot possibly cause any interference

T5D06
This question has been withdrawn.

T5D07
Which of the following may be useful in correcting a radio frequency interference problem?
A. Snap-on ferrite chokes
B. Low-pass and high-pass filters
C. Notch and band-pass filters
D. All of these answers are correct

T5D01
(C)
p 3-31

T5D02
(B)
p 3-31

T5D03
(B)
p 3-31

T5D04
(C)
p 3-31

T5D05
(A)
p 3-33

T5D06

T5D07
(D)
p 3-30

T5D08
(C)
p 3-33

T5D08
What is the proper course of action to take when a neighbor reports that your radio signals are interfering with something in his home?
A. You are not required to do anything
B. Contact the FCC to see if other interference reports have been filed
C. Check your station and make sure it meets the standards of good amateur practice
D. Change your antenna polarization from vertical to horizontal

T5D09
(D)
p 3-34

T5D09
What should you do if a "Part 15" device in your neighbor's home is causing harmful interference to your amateur station?
A. Work with your neighbor to identify the offending device
B. Politely inform your neighbor about the rules that require him to stop using the device if it causes interference
C. Check your station and make sure it meets the standards of good amateur practice
D. All of these answers are correct

T5D10
(D)
p 3-33

T5D10
What could be happening if another operator tells you he is hearing a variable high-pitched whine on the signals from your mobile transmitter?
A. Your microphone is picking up noise from an open window
B. You have the volume on your receiver set too high
C. You need to adjust your squelch control
D. The power wiring for your radio is picking up noise from the vehicle's electrical system

T5D11
(C)
p 3-4

T5D11
What may be the problem if another operator reports that your SSB signal is very garbled and breaks up?
A. You have the noise limiter turned on
B. The transmitter is too hot and needs to cool off
C. RF energy may be getting into the microphone circuit and causing feedback
D. You are operating on lower sideband

T5D12
(D)
p 4-17

T5D12
What might be the problem if you receive a report that your signal through the repeater is distorted or weak?
A. Your transmitter may be slightly off frequency
B. Your batteries may be running low
C. You could be in a bad location
D. All of these answers are correct

T5D13
(B)
p 4-33

T5D13
What is one of the reasons to use digital signals instead of analog signals to communicate with another station?
A. Digital systems are less expensive than analog systems
B. Many digital systems can automatically correct errors caused by noise and interference
C. Digital modulation circuits are much less complicated than any other types
D. All digital signals allow higher transmit power levels

SUBELEMENT T6

Communications modes and methods

3 exam questions - 3 groups

T6A - Modulation modes, descriptions and bandwidth (AM, FM, SSB) – 1 exam question

T6A01
What are phone transmissions?
A. The use of telephones to set up an amateur radio contact
B. A phone patch between amateur radio and the telephone system
C. Voice transmissions by radio
D. Placing the telephone handset near a radio transceiver's microphone and speaker to relay a telephone call

T6A02
Which of the following is a form of amplitude modulation?
A. Frequency modulation
B. Phase modulation
C. Single sideband
D. Phase shift keying

T6A03
What name is given to an amateur radio station that is used to connect other amateur stations to the Internet?
A. A gateway
B. A repeater
C. A digipeater
D. A beacon station

T6A04
Which type of voice modulation is most often used for long distance and weak signal contacts on the VHF and UHF bands?
A. FM
B. AM
C. SSB
D. PM

T6A05
Which type of modulation is most commonly used for VHF and UHF voice repeaters?
A. AM
B. SSB
C. PSK
D. FM

T6A06
Which emission type has the narrowest bandwidth?
A. FM voice
B. SSB voice
C. CW
D. Slow-scan TV

T6A01	(C) p 2-18
T6A02	(C) p 2-21
T6A03	(A) p 3-11
T6A04	(C) p 2-22
T6A05	(D) p 2-22
T6A06	(C) p 2-21

T6A07
(A)
p 2-21

T6A07
Which sideband is normally used for VHF and UHF SSB communications?
A. Upper sideband
B. Lower sideband
C. Suppressed sideband
D. Inverted sideband

T6A08
(C)
p 2-21

T6A08
What is the primary advantage of single sideband over FM for voice transmissions?
A. SSB signals are easier to tune in than FM signals
B. SSB signals are less likely to be bothered by noise interference than FM signals
C. SSB signals use much less bandwidth than FM signals
D. SSB signals have no advantages at all in comparison to other modes

T6A09
(D)
p 2-21

T6A09
What is the approximate bandwidth of a single-sideband voice signal?
A. 1 kHz
B. 2 kHz
C. Between 3 and 6 kHz
D. Between 2 and 3 kHz

T6A10
(C)
p 2-22

T6A10
What is the approximate bandwidth of a frequency-modulated voice signal?
A. Less than 500 Hz
B. About 150 kHz
C. Between 5 and 15 kHz
D. More than 30 kHz

T6A11
(B)
p 4-36

T6A11
What is the normal bandwidth required for a conventional fast-scan TV transmission using combined video and audio on the 70-centimeter band?
A. More than 10 MHz
B. About 6 MHz
C. About 3 MHz
D. About 1 MHz

T6B - Voice communications, EchoLink and IRLP – 1 exam question

T6B01
How is information transmitted between stations using Echolink?
A. APRS
B. PSK31
C. Internet
D. Atmospheric ducting

T6B01
(C)
p 4-18

T6B02
What does the abbreviation IRLP mean?
A. Internet Radio Linking Project
B. Internet Relay Language Protocol
C. International Repeater Linking Project
D. International Radio Linking Project

T6B02
(A)
p 4-18

T6B03
Who may operate on the Echolink system?
A. Only club stations
B. Any licensed amateur radio operator
C. Technician class licensed amateur radio operators only
D. Any person, licensed or not, who is registered with the Echolink system

T6B03
(B)
p 4-19

T6B04
What technology do Echolink and IRLP have in common?
A. Voice over Internet Protocol
B. Ionospheric propagation
C. AC power lines
D. PSK31

T6B04
(A)
p 4-18

T6B05
What method is used to transfer data by IRLP?
A. VHF Packet radio
B. PSK31
C. Voice over Internet Protocol
D. None of these answers is correct

T6B05
(C)
p 4-18

T6B06
What does the term IRLP describe?
A. A method of encrypting data
B. A method of linking between two or more amateur stations using the Internet
C. A low powered radio using infra-red frequencies
D. An international logging program

T6B06
(B)
p 4-18

T6B07
Which one of the following allows computer-to-radio linking for voice transmission?
A. Grid modulation
B. EchoLink
C. AMTOR
D. Multiplex

T6B07
(B)
p 4-19

T6B08
(C)
p 4-19

T6B08
What are you listening to if you hear a brief tone and then a station from Russia calling CQ on a 2-meter repeater?
A. An ionospheric band opening on VHF
B. A prohibited transmission
C. An Internet linked DX station
D. None of these answers are correct

T6B09

T6B09
This question has been withdrawn.

T6B10
(C)
p 4-14, 4-19

T6B10
Where might you find a list of active nodes using VoIP?
A. The FCC Rulebook
B. From your local emergency coordinator
C. A repeater directory or the Internet
D. The local repeater frequency coordinator

T6B11
(D)
p 4-19

T6B11
When using a portable transceiver how do you select a specific IRLP node?
A. Choose a specific CTCSS tone
B. Choose the correct DSC tone
C. Access the repeater autopatch
D. Use the keypad to transmit the IRLP node numbers

T6C – Non-voice communications - image communications, data, CW, packet, PSK31, Morse code techniques, Q signals – 1 exam question

T6C01

Which of the following is an example of a digital communications method?
A. Single sideband voice
B. Amateur television
C. FM voice
D. Packet radio

T6C01
(D)
p 4-33

T6C02

What does the term APRS mean?
A. Automatic Position Reporting System
B. Associated Public Radio Station
C. Auto Planning Radio Set-up
D. Advanced Polar Radio System

T6C02
(A)
p 4-34

T6C03

What item is required along with your normal radio for sending automatic location reports?
A. A connection to the vehicle speedometer
B. A connection to a WWV receiver
C. A connection to a broadcast FM sub-carrier receiver
D. A global positioning system receiver

T6C03
(D)
p 4-34

T6C04

What type of transmission is indicated by the term NTSC?
A. A Normal Transmission mode in Static Circuit
B. A special mode for earth satellite uplink
C. A standard fast scan color television signal
D. A frame compression scheme for TV signal

T6C04
(C)
p 4-36

T6C05

What emission mode may be used by a Technician class operator in the 219 - 220 MHz frequency range?
A. Slow-scan television
B. Point-to-point digital message forwarding
C. FM voice
D. Fast-scan television

T6C05
(B)
p 5-13

T6C06

What does the abbreviation PSK mean?
A. Pulse Shift Keying
B. Phase Shift Keying
C. Packet Short Keying
D. Phased Slide Keying

T6C06
(B)
p 4-34

T6C07

What is PSK31?
A. A high-rate data transmission mode used to transmit files
B. A method of reducing noise interference to FM signals
C. A type of television signal
D. A low-rate data transmission mode that works well in noisy conditions

T6C07
(D)
p 4-34

T6C08
(C)
p 4-13

T6C08
What sending speed is recommended when using Morse code?
A. Only speeds below five WPM
B. The highest speed your keyer will operate
C. Any speed at which you can reliably receive
D. The highest speed at which you can control the keyer

T6C09
(D)
p 4-14

T6C09
What is a practical reason for being able to copy CW when using repeaters?
A. To send and receive messages others cannot overhear
B. To conform with FCC licensing requirements
C. To decode packet radio transmissions
D. To recognize a repeater ID sent in Morse code

T6C10
(A)
p 4-4

T6C10
What is the "Q" signal used to indicate that you are receiving interference from other stations?
A. QRM
B. QRN
C. QTH
D. QSB

T6C11
(B)
p 4-4

T6C11
What is the "Q" signal used to indicate that you are changing frequency?
A. QRU
B. QSY
C. QSL
D. QRZ

SUBELEMENT T7

Special operations –
2 exam questions – 2 groups

T7A – Operating in the field, radio direction finding, radio control, contests, special event stations – 1 exam question

T7A01
What is a good thing to have when operating a hand-held transceiver away from home
A. A selection of spare parts
B. A programming cable to load new channels
C. One or more fully charged spare battery packs
D. A dummy load

T7A02
Which of these items would probably not be very useful to include in an emergency response kit
A. An external antenna and several feet of connecting cable
B. A 1500 watt output linear amplifier
C. A cable and clips for connecting your transceiver to an external battery
D. A listing of repeater frequencies and nets in your area

T7A03
How can you make the signal from a hand-held radio stronger when operating in the field?
A. Switch to VFO mode
B. Use an external antenna instead of the rubber-duck antenna
C. Stand so there is a metal building between you and other stations
D. Speak as loudly as you can

T7A04
What would be a good thing to have when operating from a location that includes lots of crowd noise?
A. A portable bullhorn
B. An encrypted radio
C. A combination headset and microphone
D. A pulse noise blanker

T7A05
What is a method used to locate sources of noise interference or jamming?
A. Echolocation
B. Doppler radar
C. Radio direction finding
D. Phase locking

T7A06
Which of these items would be the most useful for a hidden transmitter hunt?
A. Binoculars and a compass
B. A directional antenna
C. A calibrated noise bridge
D. Calibrated SWR meter

T7A01	(C) p 4-27
T7A02	(B) p 4-27
T7A03	(B) p 3-27
T7A04	(C) p 4-27
T7A05	(C) p 4-30
T7A06	(B) p 4-30

T7A07
(A)
p 4-30

T7A07
What is a popular operating activity that involves contacting as many stations as possible during a specified period of time?
A. Contesting
B. Net operations
C. Public service events
D. Simulated emergency exercises

T7A08

T7A08
This question has been withdrawn.

T7A09
(A)
p 4-5

T7A09
What is a grid locator?
A. A letter-number designator assigned to a geographic location
B. Your azimuth and elevation
C. Your UTC location
D. The 4 digits that follow your ZIP code

T7A10
(C)
p 4-30

T7A10
What is a special event station?
A. A station that sends out birthday greetings
B. A station that operates only on holidays
C. A temporary station that operates in conjunction with an activity of special significance
D. A station that broadcasts special events

T7A11
(B)
[97.215(c)]
p 4-36

T7A11
What is the maximum power allowed when transmitting telecommand signals to radio controlled models?
A. 500 milliwatts
B. 1 watt
C. 25 watts
D. 1500 watts

T7A12
(C)
[97.215(a)]
p 4-36

T7A12
What is the station identification requirement when sending commands to a radio control model using amateur frequencies?
A. Voice identification must be transmitted every 10 minutes
B. Morse code ID must be sent once per hour
C. A label indicating the licensee's call sign and address must be affixed to the transmitter
D. There is no station identification requirement for this service

T7B – Satellite operation, Doppler shift, satellite sub bands, LEO, orbit calculation, split frequency operation, operating protocols, AMSAT, ISS communications – 1 exam question

T7B01
What class of license is required to use amateur satellites?
A. Only Extra class licensees can use amateur radio satellites
B. General or higher class licensees who have a satellite operator certification
C. Only persons who are AMSAT members and who have paid their dues
D. Any amateur whose license allows them to transmit on the satellite uplink frequency

T7B01
(D)
p 4-31

T7B02
How much power should you use to transmit when using an amateur satellite?
A. The maximum power of your transmitter
B. The minimum amount of power needed to complete the contact
C. No more than half the rating of your linear amplifier
D. Never more than 1 watt

T7B02
(B)
p 4-32

T7B03
What is something you can do when using an amateur radio satellite?
A. Listen to the Space Shuttle
B. Get global positioning information
C. Make autopatch calls
D. Talk to amateur radio operators in other countries

T7B03
(D)
p 4-31

T7B04
Who may make contact with an astronaut on the International Space Station using amateur radio frequencies?
A. Only members of amateur radio clubs at NASA facilities
B. Any amateur with a Technician or higher class license
C. Only the astronaut's family members who are hams
D. You cannot talk to the ISS on amateur radio frequencies

T7B04
(B)
p 4-32

T7B05
What is a satellite beacon?
A. The primary transmit antenna on the satellite
B. An indicator light that that shows where to point your antenna
C. A reflective surface on the satellite
D. A signal that contains information about a satellite

T7B05
(D)
p 4-31

T7B06
What should you use to determine when you can access an amateur satellite?
A. A GPS receiver
B. A field strength meter
C. A telescope
D. A satellite tracking program

T7B06
(D)
p 4-31

T7B07
What is Doppler shift?
A. A change in the satellite orbit
B. A mode where the satellite receives signals on one band and transmits on another
C. A change in signal frequency caused by motion through space
D. A special digital communications mode for some satellites

T7B07
(C)
p 4-31

T7B08
(C)
p 4-32

T7B08
What is the name of the group that coordinates the building and/or launch of the largest number of amateur radio satellites?
A. NSA
B. USOC
C. AMSAT
D. FCC

T7B09
(C)
p 4-31

T7B09
What is a satellite sub-band?
A. A special frequency for talking to submarines
B. A frequency range limited to Extra Class licensees
C. A portion of a band where satellite operations are permitted
D. An obsolete term that has no meaning

T7B10
(B)
p 4-31

T7B10
What is the satellite sub-band on 70 cm?
A. 420 to 450 MHz
B. 435 to 438 MHz
C. 440 to 450 MHz
D. 432 to 433 MHz

T7B11
(C)
p 4-31

T7B11
What do the initials LEO tell you about an amateur satellite?
A. The satellite battery is in Low Energy Operation mode
B. The satellite is performing a Lunar Ejection Orbit maneuver
C. The satellite is in a Low Earth Orbit
D. The satellite uses Light Emitting Optics

SUBELEMENT T8

Emergency and Public Service Communications
3 exam questions – 3 groups

T8A - FCC declarations of an emergency, use of non-amateur equipment and frequencies, use of equipment by unlicensed persons, tactical call signs – 1 exam question

T8A01
What information is included in an FCC declaration of a temporary state of communication emergency?
A. A list of organizations authorized to use radio communications in the affected area
B. A list of amateur frequency bands to be used in the affected area
C. Any special conditions and rules to be observed during the emergency
D. An operating schedule for authorized amateur emergency stations

T8A02
Under what conditions are amateur stations allowed to communicate with stations operating in other radio services?
A. When communicating with the space shuttle
B. When specially authorized by the FCC, or in an actual emergency
C. When communicating with stations in the Citizens Radio Service
D. When a commercial broadcast station is reporting news during a natural disaster

T8A03
What should you do if you are in contact with another station and an emergency call is heard?
A. Tell the calling station that the frequency is in use
B. Direct the calling station to the nearest emergency net frequency
C. Disregard the call and continue with your contact
D. Stop your contact immediately and take the emergency call

T8A04
What are the restrictions on amateur radio communications after the FCC has declared a communications emergency?
A. The emergency declaration prohibits all communications
B. There are no restrictions if you have a special emergency certification
C. You must avoid those frequencies dedicated to supporting the emergency unless you are participating in the relief effort
D. Only military stations are allowed to use the amateur radio frequencies during an emergency

T8A05
What is one reason for using tactical call signs such as "command post" or "weather center" during an emergency?
A. They help to keep the general public informed
B. They are more efficient and help coordinate public-service communications
C. They are required by the FCC
D. They increase goodwill and sound professional

T8A01
(C)
[97.401(b)]
p 4-25

T8A02
(B)
[97.113(a)(3)]
p 4-25

T8A03
(D)
p 4-25

T8A04
(C)
p 4-25

T8A05
(B)
p 4-26

T8A06
(A)
[97.401(b)]
p 4-25

T8A06
What is legally required to restrict a frequency to emergency-only communication?
A. An FCC declaration of a communications emergency
B. Determination by the designated net manager for an emergency net
C. Authorization by an ARES/RACES emergency coordinator
D. A Congressional declaration of intent

T8A07
(D)
p 4-25

T8A07
Who has the exclusive use of a frequency if the FCC has not declared a communication emergency?
A. Any net station that has traffic
B. The station first occupying the frequency
C. Individuals passing health and welfare communications
D. No station has exclusive use in this circumstance

T8A08
(B)
p 4-25

T8A08
What should you do if you hear someone reporting an emergency?
A. Report the station to the FCC immediately
B. Assume the emergency is real and act accordingly
C. Ask the other station to move to a different frequency
D. Tell the station to call the police on the telephone

T8A09
(D)
p 4-25

T8A09
What is an appropriate way to initiate an emergency call on amateur radio?
A. Yell as loudly as you can into the microphone
B. Ask if the frequency is in use and wait for someone to give you permission to go ahead before proceeding
C. Declare a communications emergency
D. Say "Mayday, Mayday, Mayday" followed by "any station come in please" and identify your station

T8A10
(D)
p 4-25

T8A10
What are the penalties for making a false emergency call?
A. You could have your license revoked
B. You could be fined a large sum of money
C. You could be sent to prison
D. All of these answers are correct

T8A11
(B)
[97.101(c)]
p 4-24

T8A11
What type of communications has priority at all times in the Amateur Radio Service?
A. Repeater communications
B. Emergency communications
C. Simplex communications
D. Third-party communications

T8A12
(D)
[97.101(c)]
p 4-24

T8A12
When must priority be given to stations providing emergency communications?
A. Only when operating under RACES
B. Only when an emergency has been declared
C. Any time a net control station is on the air
D. At all times and on all frequencies

T8B - Preparation for emergency operations, RACES/ARES, safety of life and property, using ham radio at civic events, compensation prohibited – 1 exam question

T8B01
What can you do to be prepared for an emergency situation where your assistance might be needed?
A. Check at least twice a year to make sure you have all of your emergency response equipment and know where it is
B. Make sure you have a way to run your equipment if there is a power failure in your area
C. Participate in drills that test your ability to set up and operate in the field
D. All of these answers are correct

T8B01
(D)
p 4-27

T8B02
When may you use your amateur station to transmit a "SOS" or "MAYDAY" signal?
A. Only when you are transmitting from a ship at sea
B. Only at 15 and 30 minutes after the hour
C. When there is immediate threat to human life or property
D. When the National Weather Service has announced a weather warning

T8B02
(C)
[97.403
p 4-25

T8B03
What is the primary function of RACES in relation to emergency activities?
A. RACES organizations are restricted to serving local, state, and federal government emergency management agencies
B. RACES supports agencies like the Red Cross, Salvation Army, and National Weather Service
C. RACES supports the National Traffic System
D. RACES is a part of the National Emergency Warning System

T8B03
(A)
p 4-27

T8B04
What is the primary function of ARES in relation to emergency activities?
A. ARES organizations are restricted to serving local, state, and federal government emergency management agencies
B. ARES supports agencies like the Red Cross, Salvation Army, and National Weather Service
C. ARES groups work only with local school districts
D. ARES supports local National Guard units

T8B04
(B)
p 4-27

T8B05
What organization must you register with before you can participate in RACES activities?
A. A local amateur radio club
B. A local racing organization
C. The responsible civil defense organization
D. The Federal Communications Commission

T8B05
(C)
[97.407(a)]
p 4-27

T8B06
What is necessary before you can join an ARES group?
A. You are required to join the ARRL
B. You must have an amateur radio license
C. You must have an amateur radio license and have Red Cross CPR training
D. You must register with a civil defense organization

T8B06
(B)
p 4-27

T8B07
(D)
p 4-27

T8B07
What could be used as an alternate source of power to operate radio equipment during emergencies?
A. The battery in a car or truck
B. A bicycle generator
C. A portable solar panel
D. All of these answers are correct

T8B08
(B)
[97.403,
97.405(a),(b)]
p 4-24

T8B08
When can you use non-amateur frequencies or equipment to call for help in a situation involving immediate danger to life or property?
A. Never; your license only allows you to use the frequencies authorized to your class of license
B. In a genuine emergency you may use any means at your disposal to call for help on any frequency
C. When you have permission from the owner of the set
D. When you have permission from a police officer on the scene

T8B09
(C)
p 4-25

T8B09
Why should casual conversation between stations during a public service event be avoided?
A. Such chatter is often interesting to bystanders
B. Other listeners might overhear personal information
C. Idle chatter may interfere with important traffic
D. You might have to change batteries more often

T8B10
(B)
p 4-24

T8B10
What should you do if a reporter asks to use your amateur radio transceiver to make a news report?
A. Allow the use but give your call sign every 10 minutes
B. Advise them that the FCC prohibits such use
C. Tell them it is OK as long as you do not receive compensation
D. Tell the reporter that you must approve the material beforehand

T8B11
(C)
[97.403,
97.405(a),(b)
p 4-24

T8B11
When can you use a modified amateur radio transceiver to transmit on the local fire department frequency?
A. When you are helping the Fire Department raise money
B. Only when the Fire Department is short of regular equipment
C. In a genuine emergency you may use any means at your disposal to call for help on any frequency
D. When the local Fire Chief has given written permission

T8C - Net operations, responsibilities of the net control station, message handling, interfacing with public safety officials - 1 exam question

T8C01
Which type of traffic has the highest priority?
A. Emergency traffic
B. Priority traffic
C. Health and welfare traffic
D. Routine traffic

T8C02
What type of messages should not be transmitted over amateur radio frequencies during emergencies?
A. Requests for supplies
B. Personal information concerning victims
C. A schedule of relief operators
D. Estimates of how much longer the emergency will last

T8C03
What should you do to minimize disruptions to an emergency traffic net once you have checked in?
A. Whenever the net frequency is quiet, announce your call sign and location
B. Move 5 kHz away from the net's frequency and use high power to ask other hams to keep clear of the net frequency
C. Do not transmit on the net frequency until asked to do so by the net control station
D. Wait until the net frequency is quiet, then ask for any emergency traffic for your area

T8C04
What is one thing that must be included when passing emergency messages?
A. The call signs of all the stations passing the message
B. The name of the person originating the message
C. A status report
D. The message title

T8C05
What is one way to reduce the chances of casual listeners overhearing sensitive emergency traffic?
A. Pass messages using a non-voice mode such as packet radio or Morse code
B. Speak as rapidly as possible to reduce your on-air time
C. Spell out every word using phonetics
D. Restrict transmission of messages to the hours between midnight and 4:00 AM

T8C06
What is of primary importance for a net control station?
A. A dual-band transceiver
B. A network card
C. A strong and clear signal
D. The ability to speak several languages

T8C07
What should the net control station do if someone breaks in with emergency traffic?
A. Ask them to wait until the roll has been called
B. Stop all net activity until the emergency has been handled
C. Ask the station to call the local police and then resume normal net activities
D. Ask them to move off your net frequency immediately

T8C01
(A)
p 4-21

T8C02
(B)
p 4-25

T8C03
(C)
p 4-21

T8C04
(B)
p 4-23

T8C05
(A)
p 4-25

T8C06
(C)
p 4-21

T8C07
(B)
p 4-21

T8C08
(C)
p 4-21

T8C08
What should you do if a large scale emergency has just occurred and no net control station is available?
A. Wait until the assigned net control station comes on the air and pass your traffic when called
B. Transmit a call for help and hope someone will hear you
C. Open the emergency net immediately and ask for check-ins
D. Listen to the local NOAA weather broadcast to find out how long the emergency will last

T8C09
(D)
p 4-22

T8C09
What is the preamble of a message?
A. The first paragraph of the message text
B. The message number
C. The priority handling indicator for the message
D. The information needed to track the message as it passes through the amateur radio traffic handling system

T8C10
(A)
p 4-22

T8C10
What is meant by the term "check" in reference to a message?
A. The check is a count of the number of words in the message
B. The check is the value of a money order attached to the message
C. The check is a list of stations that have relayed the message
D. The check is a box on the message form that tells you the message was received

T8C11
(B)
p 4-22

T8C11
What is the recommended guideline for the maximum number of words to be included in the text of an emergency message?
A. 10 words
B. 25 words
C. 50 words
D. 75 words

SUBELEMENT T9

Radio waves, propagation, and antennas

3 exam questions – 3 groups

T9A - Antenna types – vertical, horizontal, concept of gain, common portable and mobile antennas, losses with short antennas, relationships between antenna length and frequency, dummy loads - 1 exam question

T9A01
What is a beam antenna?
A. An antenna built from metal I-beams
B. An antenna that transmits and receives equally well in all directions
C. An antenna that concentrates signals in one direction
D. An antenna that reverses the phase of received signals

T9A02
What is an antenna that consists of a single element mounted perpendicular to the Earth's surface?
A. A conical monopole
B. A horizontal antenna
C. A vertical antenna
D. A traveling wave antenna

T9A03
What type of antenna is a simple dipole mounted so the elements are parallel to the Earth's surface?
A. A ground wave antenna
B. A horizontal antenna
C. A rhombic antenna
D. A vertical antenna

T9A04
What is a disadvantage of the "rubber duck" antenna supplied with most hand held radio transceivers?
A. It does not transmit or receive as effectively as a full sized antenna
B. It is much more expensive than a standard antenna
C. If the rubber end cap is lost it will unravel very quickly
D. It transmits a circular polarized signal

T9A05
How does the physical size of half-wave dipole antenna change with operating frequency?
A. It becomes longer as the frequency increases
B. It must be made larger because it has to handle more power
C. It becomes shorter as the frequency increases
D. It becomes shorter as the frequency deceases

T9A06
What is the advantage of $^5/_8$ wavelength over $^1/_4$ wavelength vertical antennas?
A. They are easier to match to the feed line than other types
B. Their radiation pattern concentrates energy at lower angles
C. They pick up less noise
D. Their radiation pattern concentrates energy at higher angles

T9A01
(C)
p 3-14

T9A02
(C)
p 3-13

T9A03
(B)
p 2-23

T9A04
(A)
p 3-27

T9A05
(C)
p 2-16, 3-12

T9A06
(B)
p 3-14

T9A07
(A)
p 3-4

T9A07
What is the primary purpose of a dummy load?
A. It does not radiate interfering signals when making tests
B. It will prevent over-modulation of your transmitter
C. It keeps you from making mistakes while on the air
D. It is used for close in work to prevent overloads

T9A08
(C)
p 3-14, 3-17

T9A08
What type of antennas are the quad, Yagi, and dish?
A. Antennas invented after 1985
B. Loop antennas
C. Directional or beam antennas
D. Antennas that are not permitted for amateur radio stations

T9A09
(D)
p 3-14

T9A09
What is one type of antenna that offers good efficiency when operating mobile and can be easily installed or removed?
A. A microwave antenna
B. A quad antenna
C. A traveling wave antenna
D. A magnet mount vertical antenna

T9A10
(A)
p 3-14

T9A10
What is a good reason not to use a "rubber duck" antenna inside your car?
A. Signals can be 10 to 20 times weaker than when you are outside of the vehicle
B. RF energy trapped inside the vehicle can distort your signal
C. You might cause a fire in the vehicle upholstery
D. The SWR might increase

T9A11
(C)
p 3-13

T9A11
What is the approximate length, in inches, of a quarter-wavelength vertical antenna for 146 MHz?
A. 112 inches
B. 50 inches
C. 19 inches
D. 12 inches

T9A12
(C)
p 3-12

T9A12
What is the approximate length, in inches, of a 6-meter $^1/_2$ wavelength wire dipole antenna?
A. 6 inches
B. 50 inches
C. 112 inches
D. 236 inches

T9B – Propagation, fading, multipath distortion, reflections, radio horizon, terrain blocking, wavelength vs. penetration, antenna orientation – 1 exam question

T9B01
Why are VHF/UHF signals not normally heard over long distances?
A. They are too weak to go very far
B. FCC regulations prohibit them from going more than 50 miles
C. VHF and UHF signals are usually not reflected by the ionosphere
D. They collide with trees and shrubbery and fade out

T9B01
(C)
p 2-29

T9B02
What might be happening when we hear a VHF signal from long distances?
A. Signals are being reflected from outer space
B. Someone is playing a recording to us
C. Signals are being reflected by lightning storms in our area
D. A possible cause is sporadic E reflection from a layer in the ionosphere

T9B02
(D)
p 2-30

T9B03
What is the most likely cause of sudden bursts of tones or fragments of different conversations that interfere with VHF or UHF signals?
A. The batteries in your transceiver are failing
B. Strong signals are overloading the receiver and causing undesired signals to be heard
C. The receiver is picking up low orbit satellites
D. A nearby broadcast station is having transmitter problems

T9B03
(B)
p 3-31

T9B04
What is the radio horizon?
A. The point where radio signals between two points are blocked by the curvature of the Earth
B. The distance from the ground to a horizontally mounted antenna
C. The farthest point you can see when standing at the base of your antenna tower
D. The shortest distance between two points on the Earth's surface

T9B04
(A)
p 2-28

T9B05
What should you do if a station reports that your signals were strong just a moment ago, but now they are weak or distorted?
A. Change the batteries in your radio to a different type
B. Speak more slowly so he can understand your better
C. Ask the other operator to adjust his squelch control
D. Try moving a few feet, random reflections may be causing multi- path distortion

T9B05
(D)
p 2-28

T9B06
Why do UHF signals often work better inside of buildings than VHF signals?
A. VHF signals lose power faster over distance
B. The shorter wavelength of UHF signals allows them to more easily penetrate urban areas and buildings
C. This is incorrect; VHF works better than UHF inside buildings
D. UHF antennas are more efficient than VHF antennas

T9B06
(B)
p 2-28

T9B07
What is a good thing to remember when using your hand-held VHF or UHF radio to reach a distant repeater?
A. Speak as loudly as possible to help your signal go farther
B. Keep your transmissions short to conserve battery power
C. Keep the antenna as close to vertical as you can
D. Turn off the CTCSS tone

T9B07
(C)
p 3-28

T9B08
(B)
p 3-28

T9B08
What can happen if the antennas at opposite ends of a VHF or UHF line of sight radio link are not using the same polarization?
A. The modulation sidebands might become inverted
B. Signals could be as much as 100 times weaker
C. Signals have an echo effect on voices
D. Nothing significant will happen

T9B09
(B)
p 2-28

T9B09
What might be a way to reach a distant repeater if buildings or obstructions are blocking the direct line of sight path?
A. Change from vertical to horizontal polarization
B. Try using a directional antenna to find a path that reflects signals to the repeater
C. Ask the repeater owners to repair their receiver
D. Transmit on the repeater output frequency

T9B10
(B)
p 2-28

T9B10
What term is commonly used to describe the rapid fluttering sound sometimes heard from mobile stations that are moving while transmitting?
A. Flip-flopping
B. Picket fencing
C. Frequency shifting
D. Pulsing

T9B11
(C)
p 2-28

T9B11
Why do VHF and UHF Radio signals usually travel about a third farther than the visual line of sight distance between 2 stations?
A. Radio signals move somewhat faster than the speed of light and travel farther in the same amount of time
B. Radio waves are not blocked by dust particles
C. The Earth seems less curved to radio waves than to light
D. Radio waves are blocked by dust particles

T9C – Feedlines types, losses vs. frequency, SWR concepts, measuring SWR, matching and power transfer, weather protection, feedline failure modes – 1 exam question

T9C01
What, in general terms, is standing wave ratio (SWR)?
A. A measure of how well a load is matched to a transmitter
B. The ratio of high to low impedance in a feed line
C. The transmitter efficiency ratio
D. An indication of the quality of your station ground connection

T9C02
What reading on a SWR meter indicates a perfect impedance match between the antenna and the feed line?
A. 2 to 1
B. 1 to 3
C. 1 to 1
D. 10 to 1

T9C03
What might be indicated by erratic changes in SWR readings?
A. The transmitter is being modulated
B. A loose connection in your antenna or feedline
C. The transmitter is being over modulated
D. Interference from other stations is distorting your signal

T9C04
What is the SWR value where the protection circuits in most solid-state transmitters begin to reduce transmitter power?
A. 2 to 1
B. 1 to 2
C. 6 to 1
D. 10 to 1

T9C05
What happens to the power lost in a feed line?
A. It increases the SWR
B. It comes back into your transmitter and could cause damage
C. It is converted into heat by losses in the line
D. It can cause distortion of your signal

T9C06
What instrument other than a SWR meter could you use to determine if your feedline and antenna are properly matched?
A. Voltmeter
B. Ohmmeter
C. Iambic pentameter
D. Directional wattmeter

T9C07
What is the most common reason for failure of coaxial cables?
A. Moisture contamination
B. Gamma rays
C. End of service life
D. Overloading

T9C01
(A)
p 2-26

T9C02
(C)
p 2-26

T9C03
(B)
p 2-26

T9C04
(A)
p 2-26

T9C05
(C)
p 2-25

T9C06
(D)
p 2-26

T9C07
(A)
p 3-18

T9C08
(B)
p 2-26

T9C08
Why is it important to have a low SWR in an antenna system that uses coaxial cable feedline?
A. To reduce television interference
B. To allow the efficient transfer of power and reduce losses
C. To prolong antenna life
D. To keep your signal from changing polarization

T9C09
(C)
p 3-18

T9C09
What can happen to older coaxial cables that are exposed to weather and sunlight for several years?
A. Nothing, weather and sunlight do not affect coaxial cable
B. The cable can shrink and break
C. Losses can increase dramatically
D. It will short-circuit

T9C10
(D)
p 3-17

T9C10
Why is the outer sheath of most coaxial cables black in color?
A. It is the cheapest color to use
B. To see nicks and cracks in the cabl
C. Black cables have less loss
D. Black provides protection against ultraviolet damage

T9C11
(B)
p 2-26

T9C11
What is the impedance of the most commonly used coaxial cable in typical amateur radio installations?
A. 8 Ohms
B. 50 Ohms
C. 600 Ohms
D. 12 Ohms

T9C12
(A)
p 2-25

T9C12
Why is coaxial cable used more often than any other feed line for amateur radio antenna systems?
A. It is easy to use and requires few special installation consideration
B. It has less loss than any other type of feedline
C. It can handle more power than any other type of feedline
D. It is less expensive than any other types of line

SUBELEMENT T0

Electrical and RF Safety

3 exam questions – 3 groups

T0A – AC power circuits, hazardous voltages, fuses and circuit breakers, grounding, lightning protection, battery safety, electrical code compliance – 1 exam question

T0A01
What is a commonly accepted value for the lowest voltage that can cause a dangerous electric shock?
A. 12 volts
B. 30 volts
C. 120 volts
D. 300 volts

T0A01
(B)
p 7-2

T0A02
What is the lowest amount of electrical current flowing through the human body that is likely to cause death?
A. 10 microamperes
B. 100 milliamperes
C. 10 amperes
D. 100 amperes

T0A02
(B)
p 7-2

T0A03
What is connected to the green wire in a three-wire electrical plug?
A. Neutral
B. Hot
C. Ground
D. The white wire

T0A03
(C)
p 7-3

T0A04
What is the purpose of a fuse in an electrical circuit?
A. To make sure enough power reaches the circuit
B. To interrupt power in case of overload
C. To prevent television interference
D. To prevent shocks

T0A04
(B)
p 2-11

T0A05
What might happen if you install a 20-ampere fuse in your transceiver in the place of a 5-ampere fuse?
A. The larger fuse would better protect your transceiver from using too much current
B. The transceiver will run cooler
C. Excessive current could cause a fire
D. The transceiver would not be able to produce as much RF output

T0A05
(C)
p 2-11

T0A06
What is a good way to guard against electrical shock at your station?
A. Use 3-wire cords and plugs for all AC powered equipment
B. Connect all AC powered station equipment to a common ground
C. Use a ground-fault interrupter at each electrical outlet
D. All of these answers are correct

T0A06
(D)
p 7-3

T0A07
(C)
p 7-3

T0A07
What is the most important thing to consider when installing an emergency disconnect switch at your station?
A. It must always be as near to the operator as possible
B. It must always be as far away from the operator as possible
C. Everyone should know where it is and how to use it
D. It should be installed in a metal box to prevent tampering

T0A08
(D)
p 7-6

T0A08
What precautions should be taken when a lightning storm is expected?
A. Disconnect the antenna cables from your station and move them away from your radio equipment
B. Unplug all power cords from AC outlets
C. Stop using your radio equipment and move to another room until the storm passes
D. All of these answers are correct

T0A09
(C)
p 4-27

T0A09
What is one way to recharge a 12-volt battery if the commercial power is out?
A. You cannot recharge a battery unless the power is back on
B. Add water to the battery
C. Connect the battery to a car's battery and run the engine
D. Take your battery to the utility company for a recharge

T0A10
(D)
p 3-26

T0A10
What kind of hazard is presented by a conventional 12-volt storage battery?
A. It contains dangerous acid that can spill and cause injury
B. Short circuits can damage wiring and possibly cause a fire
C. Explosive gas can collect if not properly vented
D. All of these answers are correct

T0A11
(A)
p 3-26

T0A11
What can happen if a storage battery is charged or discharged too quickly?
A. The battery could overheat and give off dangerous gas or explode
B. The terminal voltage will oscillate rapidly
C. The warranty will be voided
D. The voltage will be reversed

T0A12
(C)
p 7-6

T0A12
What is the most important reason to have a lightning protection system for your amateur radio station?
A. Lower insurance rates
B. Improved reception
C. Fire prevention
D. Noise reduction

T0A13
(D)
p 7-2

T0A13
What kind of hazard might exist in a power supply when it is turned off and disconnected?
A. Static electricity could damage the grounding system
B. Circulating currents inside the transformer might cause damage
C. The fuse might blow if you remove the cover
D. You might receive an electric shock from stored charge in large capacitors

T0B – Antenna installation, tower safety, overhead power lines – 1 exam question

T0B01

Why should you wear a hard hat and safety glasses if you are on the ground helping someone work on an antenna tower?

A. It is required by FCC rules

B. To keep RF energy away from your head during antenna testing

C. To protect your head and eyes in case something accidentally falls from the tower

D. It is required by the electrical code

T0B01
(C)
p 7-15

T0B02

What is a good precaution to observe before climbing an antenna tower?

A. Turn on all radio transmitters

B. Remove all tower grounding connections

C. Put on your safety belt and safety glasses

D. Inform the FAA and the FCC that you are working on a tower

T0B02
(C)
p 7-15

T0B03

What should you do before you climb a tower?

A. Arrange for a helper or observer

B. Inspect the tower for damage or loose hardware

C. Make sure there are no electrical storms nearby

D. All of these answers are correct

T0B03
(D)
p 7-15

T0B04

What is an important consideration when putting up an antenna?

A. Carefully tune it for a low SWR

B. Make sure people cannot accidentally come into contact with it

C. Make sure you discard all packing material in a safe place

D. Make sure birds can see it so they don't fly into it

T0B04
(B)
p 7-13

T0B05

What must be considered when erecting an antenna near an airport?

A. The maximum allowed height with regard to nearby airports

B. The possibility of interference to aircraft radios

C. The radiation angle of the signals it produces

D. The polarization of signal to be radiated

T0B05
(A)
[97.15(A)]
p 7-13

T0B06

What is the most important safety precaution to observe when putting up an antenna tower?

A. Install steps on the tower for safe climbing

B. Insulate the base of the tower to avoid lightning strikes

C. Ground the base of the tower to prevent lightning strikes

D. Look for and stay clear of any overhead electrical wires

T0B06
(D)
p 7-14

T0B07

How should the guy wires for an antenna tower be installed?

A. So each guy wire anchor point has an even number of wires

B. So that no guy wire is more than 25 feet long

C. Each guy wire must be pulled as tight as possible

D. In accordance with the tower manufacturer's instructions

T0B07
(D)
p 7-14

T0B08
(D)
p 7-14

T0B08
What is a safe distance from a power line to allow when installing an antenna?
A. Half the width of your property unless the wires are at least 23 feet high
B. 12.5 feet in most metropolitan areas
C. 36 meters plus $\frac{1}{2}$ wavelength at the operating frequency
D. So that if the antenna falls unexpectedly, no part of it can come closer than 10 feet to the power wires

T0B09
(D)
p 7-15

T0B09
What is the most important safety rule to remember when using a crank-up tower?
A. This type of tower must never be painted
B. Crank up towers must be raised and lowered frequently to keep them properly lubricated
C. Winch cables must be specially rated for use on this type of tower
D. A crank-up tower should never be climbed unless it is in the fully lowered position

T0B10
(C)
p 3-23

T0B10
Why is stainless steel hardware used on many antennas instead of other metals?
A. Stainless steel is a better electrical conductor
B. Stainless steel weighs less than other metals
C. Stainless steel parts are much less likely to corrode
D. Stainless steel costs less than other metals

T0B11
(C)
p 7-14

T0B11
What is considered to be an adequate ground for a tower?
A. A single 4 foot ground rod, driven into the earth no more than 12 inches from the base
B. A screen of 120 radial wires
C. Separate 8 foot long ground rods for each tower leg, bonded to the tower and each other
D. A connection between the tower base and a cold water pipe

T0C - RF hazards, radiation exposure, RF heating hazards, proximity to antennas, recognized safe power levels, hand held safety, exposure to others - 1 exam question

T0C01

What type of radiation are VHF and UHF radio signals?

A. Gamma radiation
B. Ionizing radiation
C. Alpha radiation
D. Non-ionizing radiation

T0C02

When can radio waves cause injury to the human body?

A. Only when the frequency is below 30 MHz
B. Only if the combination of signal strength and frequency cause excessive power to be absorbed
C. Only when the frequency is greater than 30 MHz
D. Only when transmitter power exceeds 50 watts

T0C03

What is the maximum power level that an amateur radio station may use at VHF frequencies before an RF exposure evaluation is required?

A. 1500 watts PEP transmitter output
B. 1 watt forward power
C. 50 watts PEP at the antenna
D. 50 watts PEP reflected power

T0C04

What factors affect the RF exposure of people near an amateur transmitter?

A. Frequency and power level of the RF field
B. Distance from the antenna to a person
C. Radiation pattern of the antenna
D. All of these answers are correct

T0C05

Why must the frequency of an RF source be considered when evaluating RF radiation exposure?

A. Lower frequency RF fields have more energy than higher frequency fields
B. Lower frequency RF fields do not penetrate the human body
C. Higher frequency RF fields are transient in nature and do not affect the human body
D. The human body absorbs more RF energy at some frequencies than others

T0C06

How can you determine that your station complies with FCC RF exposure regulations?

A. By calculation based on FCC OET Bulletin 65
B. By calculation based on computer modeling
C. By measurement of field strength using calibrated equipment
D. All of these choices are correct

T0C07

What could happen if a person accidentally touched your antenna while you were transmitting?

A. Touching the antenna could cause television interference
B. They might receive a painful RF burn injury
C. They would be able to hear what you are saying
D. Nothing

T0C01	(D) p 7-7
T0C02	(B) p 7-7
T0C03	(C) [97.13(C)(1)] p 7-11
T0C04	(D) p 7-11
T0C05	(D) p 7-8
T0C06	(D) [97.13(c)(1)] p 7-10
T0C07	(B) p 7-7

T0C08
(D)
p 7-11

T0C08

What action might amateur operators take to prevent exposure to RF radiation in excess of FCC supplied limits?
A. Alter antenna patterns
B. Relocate antennas
C. Change station parameters such as frequency or power
D. All of these answers are correct

T0C09
(B)
p 7-11

T0C09

How can you make sure your station stays in compliance with RF safety regulations?
A. Compliance is not necessary
B. By re-evaluating the station whenever an item of equipment is changed
C. By making sure your antennas have a low SWR
D. By installing a low pass filte

T0C10
(A)
p 7-7

T0C10

Which of the following units of measurement is used to measure RF radiation exposure?
A. Milliwatts per square centimeter
B. Megohms per square meter
C. Microfarads per foot
D. Megahertz per second

T0C11
(A)
p 7-9

T0C11

Why is duty cycle one of the factors used to determine safe RF radiation exposure levels?
A. It takes into account the amount of time the transmitter is operating
B. It takes into account the transmitter power supply rating
C. It takes into account the antenna feed line loss
D. It takes into account the thermal effects of the final amplifier

About the ARRL

The seed for Amateur Radio was planted in the 1890s, when Guglielmo Marconi began his experiments in wireless telegraphy. Soon he was joined by dozens, then hundreds, of others who were enthusiastic about sending and receiving messages through the air—some with a commercial interest, but others solely out of a love for this new communications medium. The United States government began licensing Amateur Radio operators in 1912.

By 1914, there were thousands of Amateur Radio operators—hams—in the United States. Hiram Percy Maxim, a leading Hartford, Connecticut inventor and industrialist, saw the need for an organization to band together this fledgling group of radio experimenters. In May 1914 he founded the American Radio Relay League (ARRL) to meet that need.

Today ARRL, with approximately 150,000 members, is the largest organization of radio amateurs in the United States. The ARRL is a not-for-profit organization that:
• promotes interest in Amateur Radio communications and experimentation
• represents US radio amateurs in legislative matters, and
• maintains fraternalism and a high standard of conduct among Amateur Radio operators.

At ARRL headquarters in the Hartford suburb of Newington, the staff helps serve the needs of members. ARRL is also International Secretariat for the International Amateur Radio Union, which is made up of similar societies in 150 countries around the world.

ARRL publishes the monthly journal *QST*, as well as newsletters and many publications covering all aspects of Amateur Radio. Its headquarters station, W1AW, transmits bulletins of interest to radio amateurs and Morse code practice sessions. The ARRL also coordinates an extensive field organization, which includes volunteers who provide technical information and other support services for radio amateurs as well as communications for public-service activities. In addition, ARRL represents US amateurs with the Federal Communications Commission and other government agencies in the US and abroad.

Membership in ARRL means much more than receiving *QST* each month. In addition to the services already described, ARRL offers membership services on a personal level, such as the ARRL Volunteer Examiner Coordinator Program and a QSL bureau.

Full ARRL membership (available only to licensed radio amateurs) gives you a voice in how the affairs of the organization are governed. ARRL policy is set by a Board of Directors (one from each of 15 Divisions). Each year, one-third of the ARRL Board of Directors stands for election by the full members they represent. The day-to-day operation of ARRL HQ is managed by an Executive Vice President and his staff.

No matter what aspect of Amateur Radio attracts you, ARRL membership is relevant and important. There would be no Amateur Radio as we know it today were it not for the ARRL. We would be happy to welcome you as a member! (An Amateur Radio license is not required for Associate Membership.) For more information about ARRL and answers to any questions you may have about Amateur Radio, write or call:

ARRL—The national association for Amateur Radio
225 Main Street
Newington CT 06111-1494
Voice: 860-594-0200
   Fax: 860-594-0259
   E-mail: **hq@arrl.org**
   Internet: **www.arrl.org/**

Prospective new amateurs call (toll-free):
800-32-NEW HAM (800-326-3942)
You can also contact us via e-mail at **newham@arrl.org**
or check out *ARRLWeb* at **www.arrl.org/**

Index

PL™ is a registered trademark of the Motorola Corporation.

Notes

Notes

Notes

Notes

Notes

Notes

Notes

Notes

Notes

Notes

MFJ *Deluxe* Antenna Tuner

Tune your antenna for minimum SWR!
Works 1.8 to 30 MHz on dipoles, verticals, inverted vees, random wires, beams, mobile whips, shortwave receiving antennas . . . Use coax, random wire, balanced lines. Has heavy duty 4:1 balun for balanced lines.

Custom designed inductor switch, 1000 volt tuning capacitors, *Teflon*(R) insulating

MFJ-949E $169.95 washers and proper L/C ratio gives you *arc-free* no worries operation with up to 300 Watts PEP transceiver input power.

The MFJ-949E inductor switch was *custom* designed to withstand the extremely high RF voltages and currents that are developed in your tuner.

Antenna switch lets you select two coax fed antennas, random wire/balanced line or dummy load through your MFJ-949E or direct to your transceiver.

Full size 3-inch lighted Cross-Needle Meter. Lets you easily read SWR, *peak* or average forward and reflected power simul-taneously. Has 300 Watt or 30 Watt ranges.

MFJ's *QRM-Free PreTune*™ lets you pre-tune your MFJ-949E *off-the-air* into its built-in dummy load! Makes tuning your actual antenna faster and easier.

Full size built-in non-inductive 50 Ohm dummy load, scratch-proof Lexan multi-colored front panel, 10⅝x3½x7 in. Superior cabinet construction and more!

MFJ HF thru 6 Meter Antenna Tuner
Covers 1.8-30 MHz *and* 6 Meters. *Lighted* Cross-Needle SWR/Wattmeter. Lamp and RF bypass switches. 300 Watts xcvr input. 8wx2Hx6D inches.

MFJ-945E $119.95

MFJ Switching Power Supplies

Tiny 3.7 lb. switching power supply powers any transceiver! 25 Amps maximum or 22 Amps continuous. 5¾Wx4½Hx6D in. *No RF Hash!* Has Over Voltage and Over Current protection circuits. Front-panel has easy access heavy duty 5-way binding posts and convenient cigarette lighter socket. Variable 9 to 15 Volts DC. Works anywhere in the world, 85 to 135 VAC or 170 to 260 VAC. Replaceable fuse. 35 mV peak-to-peak ripple under 25 amp full load. Load regulation is better than 1.5% under full load.

MFJ-4225MV $149.95

MFJ-4125, $109.95. Super tiny 5½Wx2½Hx5¾D inches. Like MFJ-4225MV. No meters or cigarette lighter socket.

MFJ 1.8-170 MHz SWRAnalyzer

The world's most popular antenna analyzer gives you a complete picture of your antenna performance 1.8 to 170 MHz. Super easy-to-use -- makes tuning your antenna quick and easy. Read antenna SWR, complex impedance, return loss, reflection coefficient. Determine velocity factor, coax cable loss in dB, length of coax and distance to a short or open in feet. Read inductance in uH and capacitance in pF at RF frequencies. Large easy-to-read two line LCD screen and side-by-side meters clearly display your information. Built-in frequency counter, Ni-Cad charger circuit, battery saver, low battery warning and smooth reduction drive tuning.

MFJ-259B $279.95

MFJ 6-Band Vertical Antenna

Work lots of DX on 6 bands: 40, 20, 15, 10, 6, 2 Meters. No radials or ground needed!

MFJ-1796 $219.95

Only 12 feet high, tiny 24 inch footprint! Mount anywhere -- ground level to tower top -- apartments, small lots. Goes together in an afternoon.

Efficient end-loading, no lossy traps. Entire length is always radiating. Full size halfwave on 2/6 Meters.

Automatic bandswitching, low radiation angle, omni-directional, handles 1500 Watts PEP.

High power air-wound choke balun eliminates feedline radiation. No tuning interaction.

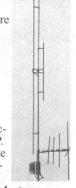

MFJ Hamshack Essentials

MFJ-1702C, $29.95. 2 position antenna coax switch. New center ground, lightning arrestor. 2.5 kW PEP, 1 kW CW. Loss below .2 dB. 50 dB isolation @ 450 MHz. 50 Ohms. 3x2x3".

MFJ-1704, $69.95. Like MFJ-1702C. 4 positions. 6¼x 4¼x1¼ in. 50 dB isolation at 500 MHz. Center Ground/Lightning arrestor.

MFJ-281, $14.95. *Communications* speaker makes copying easier! Enhances speech and CW, improves intelligibility, reduces noise, static, hum.

MFJ-862, $59.95. Covers 144, 220, 440 MHz. *Cross-Needle Meter* 30/300 Watts forward, 6/60 Watts reflected ranges. Lighted meter, single sensor.

MFJ-108B, $21.95. *Dual* 24/12 hour-clock. Read UTC and local time simultaneously. Huge high-contrast 5/8 inch LCD, brushed aluminum frame. Includes long-life battery. 4½Wx1Dx2H inches.

MFJ-260C, $39.95. Dry 300 Watt HF/VHF dummy load. SWR 1.1: 1 to 30 MHz; 1.5:1, 30-650 MHz. SO-239. 2¼x2¼x7 in.

MFJ All-Band G5RV Antenna

Covers all bands, 160-10 Meters with antenna tuner. 102 feet long, shorter than 80 Meter dipole. Use as inverted vee or sloper to be more compact. Use on 160 Meters as Marconi with tuner and ground. Handles full legal limit power. Add coax and some rope and you're *on the air!*

MFJ-1778 $44.95

MFJ-1778M, $39.95. MFJ *G5RV Junior* covers 10 through 40 Meters with tuner, handles 1500 Watts, 52 feet long. *New!*

MFJ . . . For All Your CW Needs . . .

MFJ-461, $89.95. Place this tiny MFJ Morse Code Reader near your receiver's speaker . . . Watch CW turn into solid text messages as they scroll across an LCD display.

MFJ-418, $89.95. Learn Morse code anywhere with MFJ's pocket tutor! Follows ARRL/VEC format. Go from zero code speed to CW Pro! Use earphones/built-in speaker.

MFJ-557, $34.95. Adjustable Morse key, code practice oscillator and speaker mounted on heavy steel base -- stays put. Earphone jack, tone, volume controls.

MFJ-464, $189.95. *Morse Code Reader* with built-in *Memory Keyer*. Send/receive 5-99 WPM. Weight and sidetone controls. 190 character type-ahead buffer. Use keyboard or iambic paddle.

MFJ-564, $69.95. Chrome. Deluxe MFJ *Iambic Paddles*™. Tension and contact spacing adjustments. Heavy precision machined base, non-skid feet. Self-adjusting bearings. MFJ-564B (black).

MFJ VHF/UHF Headquarters

MFJ-1717, $21.95. World's best-selling dual band *HT antenna*! 144/ 440 MHz flexible duck antenna is thin, super strong and flexible! Choose BNC or SMA.

MFJ-1724B, $19.95. Dual band 144/440 MHz *magnet mount mobile* antenna. 19" radiator, 300W, 3½" super strong magnet, *free* BNC adapter, 15' coax.

MFJ-1750, $29.95. 2 Meter, 300 Watt 5/8 wave *maximum gain* ground plane *base* antenna. Easy to put up, U-bolt included, strong lightweight aluminum.

MFJ-295, $15.95. Tiny top-quality speaker mic for tiny HTs. Earphone jack, swivel lapel clip, 3½ foot coiled cord. Black. Specify Yaesu, Icom, Kenwood, others.

take it with you?

DIGITAL

IC-91AD**
ANALOG/DIGITAL 2M/70CM

2M/70CM @ 5W I Wide Band RX 0.495
- 999.99MHz* with Dual Watch I 1300
Alphanumeric Memories I Li-ion Power I
Digital Voice & Data I VV/UU/VU

DIGITAL

IC-91A**
ANALOG/DIGITAL 2M/70CM

2M/70CM @ 5W I Wide Band RX 0.495
- 999.99MHz* with Dual Watch I 1300
Alphanumeric Memories I Li-ion Power I
Optional Digital Voice & Data I VV/UU/VU

DIGITAL

IC-V82
ANALOG/DIGITAL 2M

7 Watt Output I 100 Alphanumeric
Memories I CTCSS & DTCS Encode/Decode
with Tone Scan I Optional Digital Voice &
Data I Optional Callsign Squelch

DIGITAL

IC-U82
ANALOG/DIGITAL 70CM

5 Watt Output I 100 Alphanumeric
Memories I CTCSS & DTCS Encode/Decode
with Tone Scan I Optional Digital Voice &
Data I Optional Callsign Squelch

IC-P7A
MINI 2M/70CM

2M @ 1.5W, 70CM @ 1W I Wide
Band RX 0.495 - 999.99MHz* I 1000
Alphanumeric Memories I Up To 20 Hours
of Operating Time I Rapid Charger

IC-T90A
COMPACT 6M/2M/70CM

5 Watt Output I Wide Band RX .495
- 999.99MHz* I 500 Alphanumeric
Memories I 1300 mAh Li-Ion Battery I CTCSS
& DTCS Encode/Decode w/Tone Scan

IC-V8 SPORT
COMMERCIAL GRADE 2M

AA Alkaline Power I CTCSS & DTCS
Encode/Decode w/Tone Scan I 107
Alphanumeric Memories I Fast Scanning
- 40 Channels/Second

IC-T7H
EASY TO USE 2M/70CM

5 - 6 Watts Depending on Battery Pack I RX:
118-174, 400-470MHz I 70 Alphanumeric
Memory Channels I CTCSS Encode/Decode
with Tone Scan I Mil Spec

See our full product lineup at your authorized Icom dealer!

8432

12 STORE BUYING POWER

HAM RADIO OUTLET

DISCOVER THE POWER OF DSP WITH ICOM!

IC-7800 All Mode Transceiver

- 160-6M @ 200W • Four 32 bit IF-DSPs+ 24 bit AD/DA converters • Two completely independent receivers • +40dBm 3rd order intercept point • Now with 3rd roofing filter

IC-756PROIII All Mode Transceiver

- 160-6M • 100W • Adjustable SSB TX bandwidth • Digital voice recorder • Auto antenna tuner • RX: 30 kHz to 60 MHz • Quiet, triple-conversion receiver • 32 bit IF-DSP • Low IMD roofing filter • 8 Channel RTTY TX memory • Digital twin passband tuning • Auto or manual-adjust notch with 70 dB attenuation

IC-746PRO All Mode 160M-2M

- 160-2M* @ 100W • 32 bit IF-DSP+ 24 bit AD/DA converter • Selectable IF filter shapes for SSB & CW • Enhanced Rx performance

IC-718 HF Transceiver

- 160-10M* @ 100W • 12V Operation • Simple to Use • CW Keyer Built-in • One Touch Band Switching • Direct frequency input • VOX Built-in • Band stacking register • IF shift • 101 memories

IC-706MKIIG All Mode Transceiver

- Proven Performance • 160-10M*/6M/2M/70CM • All mode w/DSP • HF/6M @ 100W, 2M @ 50W, 440 MHz @ 20W • CTCSS encode/decode w/tone scan • Auto repeater • 107 alphanumeric memories

IC-7000 All Mode Transceiver

- **For the love of ham radio!** • 160 - 10M, 6M @ 100W (40W AM), 2M @ 50W (20W AM), 70CM @ 35W (14W AM), all continually adjustable • 2x DSP • Digital IF filters • Digital voice recorder • 2.5" color TFT display • 503 memory channels • Remote control mic

IC-2720H Dual Band Mobile

- 2M/70CM • VV/UU/VU • Wide band RX inc. air & weather bands • Dynamic Memory Scan (DMS) • CTCSS/DTCS encode/decode w/tone scan • Independent controls for each band • DTMF Encode • 212 memory channels • Remote Mount Kit Inc.

IC-208H Dual Band Mobile

- 55 watts VHF (2M), 50 watts UHF (70CM) • Wide band RX (Cellular blocked on US versions) • 500 alphanumeric memories • CTCSS/DTCS encode/decode w/tone scan • Detachable remote head • DMS w/linked banks

IC-V8000 2M Mobile Transceiver

- 75 watts • Dynamic Memory Scan (DMS) • CTCSS/DCS encode/decode w/tone scan • Weather alert • Weather channel scan • 200 alphanumeric memories

*Except 60M Band. **Cellular blocked, unblocked OK to FCC approved users. Please check with HRO for details or restrictions on any offers or promotions.
*1 For shock & vibration. *2 When connected to an external GPS ©2006 Icom America Inc. The Icom logo is a registered trademark of Icom Inc. 8439

ICOM®

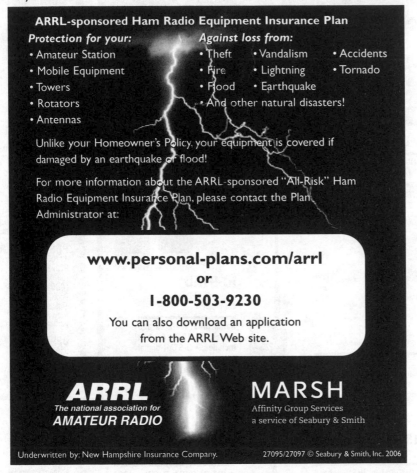

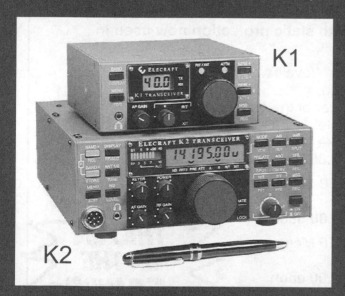

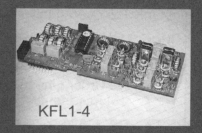

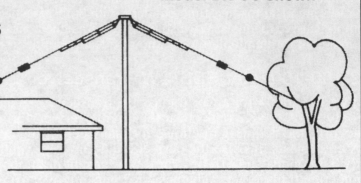

TS-480

The Perfect Remote Base Transceiver

Straight Out of the Box!

TS-480HX
HF/50MHz All-Mode Transceiver

- The perfect internet base transceiver - straight out of the box!
- Easy to operate
- The size makes it great for base, mobile or portable operation.
- Free VoIP/Control software downloads at Kenwoodusa.com.
- Incredible RX specifications.

KENWOOD U.S.A. CORPORATION
Communications Sector Headquarters
3975 Johns Creek Court, Suite 300, Suwanee, GA 30024-1265
Customer Support/Distribution
P.O. Box 22745, 2201 East Dominguez St., Long Beach, CA 90801-5745
Customer Support: (310) 639-4200   Fax: (310) 537-8235

INTERNET
Kenwood News & Products
http://www.kenwoodusa.com
ADS#10006

ISO9001 Registered
Communications Equipment Division
Kenwood Corporation
ISO9001 certification

Circuit Specialists, Inc.

Providing Electronic Equipment & Supplies Since 1971

Soldering Stations with Ceramic Element & Separate Solder Stand

$34.95

- Ceramic heating element for more accurate temp control
- Temp control knob in F(392° to 896°) & C(200° to 489°)
- 3-prong grounded power cord/static safe tip
- Separate heavy duty iron stand
- Replaceable iron/easy disconnect
- Extra tips etc. shown at web site

Item # **CSI-STATION1A** *Rapid Heat Up!*

Also Available w/ Digital Display & MicroProcessor Controller

Item # **CSI-STATION2A** **$49.95**

Details at Web Site

13.8V DC Regulated Linear Power Supplies

Very durable and rugged regulated power supplies that provide 13.8 Volts and substantial amounts of current so that they deliver real world power when you need it most! Choose from three models; the **CSI1862** with 6 Amps, the **CSI-1865** with a robust 20 Amps or the **CSI-1869** with a beefy 40 Amps at your disposal.

Item # **CSI1862** 13.8VDC / 6Amp **$24.95**

Item # **CSI-1865** 13.8VDC / 20Amp **$69.00**

Item # **CSI-1869** 13.8VDC / 40Amp **$119.00**

Details at Web Site

Soldering Stations for Lead Free Solder

Compatible with all lead-free alloy solder and standard solder. Excellent thermal recovery without a large increase in tip temperature. Utilize an integrated ceramic heater, sensor, control circuit and tip for greater efficiency, along with a highly dependable 24V output transformer. The effortless replacement of soldering tips makes for quick changes and the optional shutdown setting turns the unit off after 30 min. of idle time. Various tips are available at our Web Site!

- Power Consumption: 70W
- Output Voltage: 24VAC
- Temperature Range: 200-480°C/ 392-896°F

Item # **CSI-2901** **$59.00**

Also Available w/ Digital Display

Item # **CSI-2900**

$69.00

Details at Web Site

---

2 Amp Multi-Output Power Supply

This unit is switchable and provides regulated outputs of 3V, 4.5V, 6V, 7.5V, 9V & 12V. All outputs provide 2 Amps of power. Fuse protected. Smart grey plastic enclosure with on/off switch & red & black output jacks.

Item # **PS-28** **$19.95**

Details at Web Site

Protek
20MHz Dual Trace Oscilloscope w/Alt-Mag

- 20MHz Bandwidth
- Alt-Mag sweep for simultaneous display of main and X5 magnified trace
- 1mV/Div Vertical sensitivity
- Alternate trigger for a stable display of unrelated signals

Item # **PROTEK 6502** **$279.00**

Details at Web Site

Auto Ranging DMM

- DCV: 1000V
- ACV: 750V
- Amps: 20A AC & DC
- Frequency: (1KHz to 300KHz)
- Resistance: 30Ohm
- Temperature: 0° to 1832°
- Auto Ranging

Includes: Analog Bar Graph, Auto-Ranging, Data Hold, Temperature, Frequency Test, Continuity Test & more!

$29.95

Item # **CSI9303**

Details at Web Site

Bag O' LEDs

LED's have long leads, have a maximum rating of 5V, up to 35mA, and have a standard brightness. **Come in bags of 100**.

Item #		100pc:
BAG-RED5MM	5mm	$1.50
BAG-GREEN5MM	5mm	$1.50
BAG-YELLOW5MM	5mm	$2.00
BAG-RED3MM	3mm	$1.50
BAG-GREEN3MM	3mm	$1.50
BAG-YELLOW3MM	3mm	$2.00

Factory Firsts!

Details at Web Site

Protek
Dual Trace 100MHz Oscilloscope

- Four traces may be simultaneously displayed in ALT-sweep
- Five Vertical Modes Ch1,Ch2, Dual, Add and Subtract
- Bright 6" CRT with an internal graticule
- 12 kv acceleration voltage
- Sweep speeds to 2nS/Div.

Item # **PROTEK 6510** **$519.00**

Details at Web Site

Test Equipment, Soldering Equipment, Semiconductor Devices, Power Supplies, and much much more!

SEE OUR WEB SITE: www.circuitspecialists.com for our latest offers, and to check on stock and updated pricing on thousands of items.

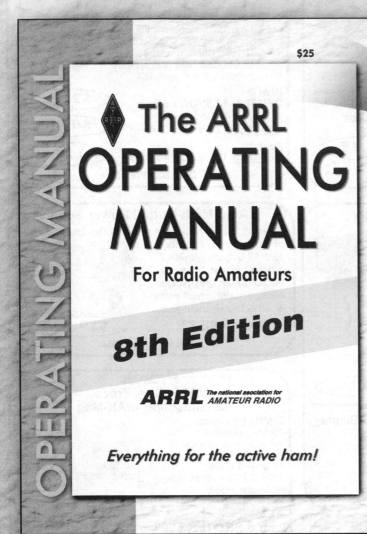

FEEDBACK

Please use this form to give us your comments on this book and what you'd like to see in future editions, or e-mail us at **pubsfdbk@arrl.org** (publications feedback). If you use e-mail, please include your name, call, e-mail address and the book title, edition and printing in the body of your message. Also indicate whether or not you are an ARRL member.

Please check the box that best answers these questions:

How well did this book prepare you for your exam? ☐ Very well ☐ Fairly well ☐ Not very well

How well did this book teach you about ham radio? ☐ Very well ☐ Fairly well ☐ Not very well

Which exam did you take (or will you be taking)? ☐ Technician ☐ Technician with code Did you pass? ☐ Yes ☐ No

If you checked Technician: Do you expect to learn Morse code at some point? ☐ Yes ☐ No

Which operating modes do you plan to use first?

☐ SSB ☐ FM ☐ Packet ☐ RTTY ☐ Image ☐ Morse code ☐ Other _____

Where did you purchase this book? ☐ From ARRL directly ☐ From an ARRL dealer

Is there a dealer who carries ARRL publications within:

☐ 5 miles ☐ 15 miles ☐ 30 miles of your location? ☐ Not sure.

Name _____ ARRL member? ☐ Yes ☐ No

_____ Call Sign _____

Address _____

City, State/Province, ZIP/Postal Code _____

Daytime Phone ( ) _____ Age _____

If licensed, how long? _____

Other hobbies _____

_____

Occupation _____

E-mail _____

For ARRL use only	HRLM
Edition	1 2 3 4 5 6 7 8 9 10 11 12
Printing	2 3 4 5 6 7 8 9 10 11 12

From _____

_____

_____

EDITOR, THE ARRL HAM RADIO LICENSE MANUAL
ARRL—THE NATIONAL ASSOCIATION FOR AMATEUR RADIO
225 MAIN STREET
NEWINGTON CT 06111-1494

please fold and tape